DARWIN'S FOSSILS

THE COLLECTION THAT SHAPED THE THEORY OF EVOLUTION

ADRIAN LISTER

Smithsonian Books

Washington, DC

To EAL and LLJ, with love.

Published in Great Britain by the Natural History Museum, Cromwell Rd, London SW7 5BD
Copyedited by Celia Coyne
Designed by Mercer Design, London
Reproduction by Saxon Digital Services

Published in North America by Smithsonian Books

This book may be purchased for educational, business, or sales promotional use. For information, please write: Special Markets Department, Smithsonian Books, P.O. Box 37012, MRC 513, Washington, DC 20013

ISBN 978-1-58834-617-9

Library of Congress Cataloging-in-Publication Data
Names: Lister, Adrian, author. | Natural History Museum (London, England)
Title: Darwin's fossils : the collection that shaped the theory of evolution
 / Adrian Lister.
Description: Washington, DC : Smithsonian Books, c2018. | "Copyrighted by the
 Trustees of the Natural History Museum, London." | Includes
 bibliographical references and index.
Identifiers: LCCN 2017046415 | ISBN 9781588346179 (pbk.)
Subjects: LCSH: Darwin, Charles, 1809-1882. | Evolution (Biology) |
 Fossils--Catalogs and collections. | Beagle Expedition (1831-1836)
Classification: LCC QH366.2 .L57 2018 | DDC 576.8/2--dc23 LC record available at
https://lccn.loc.gov/2017046415

Printed in China by Toppan Leefung Printing Limited, not at government expense
22 21 20 19 18   5 4 3 2 1

CONTENTS

~

AUTHOR'S NOTE

FOSSIL-COLLECTING WAS ONE OF DARWIN'S key preoccupations during the *Beagle* voyage, and the discoveries that he made became one of the principal lines of evidence that led him to embrace the reality of evolution. As such, his fossil specimens have tremendous significance in the history of science. We are fortunate indeed that most of these fossils survive in museum collections, many of them at the Natural History Museum in London. Yet despite specialist research on individual groups, these collections have rarely been considered as a whole, and important parts of them have barely been studied since the 19th century.

This book places Darwin's *Beagle* fossils centre stage. Almost all the illustrated specimens are those collected by Darwin himself, or in some cases by his shipmates on his behalf, during the course of the voyage. One of the great pleasures of writing this book has been the unearthing, with the help of curators responsible for the collections, of a number of long-forgotten Darwin specimens. In several cases species or larger groups of organisms have been recognised among Darwin's material for the first time.

Complementing the specimens themselves are Darwin's notebooks, diaries and letters from the voyage. With the recent accessibility of these documents it has been possible in many cases to trace current specimens back to Darwin's original description of their discovery, sometimes written on the spot. These original sources have been extensively quoted – they have a freshness that conveys Darwin's excitement and immediate thinking on making new discoveries. In his subsequent writings we follow the maturation of his views and their incorporation into a larger scheme.

In *Darwin's Fossils*, each chapter or section ends with a summary of our current ideas on the fossils and their interpretation. Here I rely very largely on the work of specialist researchers, through their published works or by kindly sharing their knowledge and insights with me. I have used the word 'evolution' throughout the text, even though it was not understood in its modern sense during the early part of Darwin's career when the term 'transmutation' signified the gradual transformation of species. By the 'theory of evolution' I refer simply to the acceptance of evolution as a historical reality, rather than Darwin's theory for its mechanism, natural selection, which he subsequently developed to explain the process.

This book could not have been written in anything like its present form without the help of two groups of people to whom I offer my sincerest thanks. The first are my colleagues at the Natural History Museum, each of whom knows far more than I do about their specialist fossil groups, and who have helped me with access to specimens, identifications, and detailed information. It is a great pleasure in this respect to thank Paul Barrett, Emma Bernard, Jill Darrell, Greg Edgecombe, Tim Ewin, Peta Hayes, Zoë Hughes, Zerina Johanson, Paul Kenrick, Claire Mellish, Noel Morris, Martha Richter, Brian Rosen, Consuelo Sendino, Chris Stringer, Paul Taylor and Jon Todd. A special thank you to Pip Brewer, fellow-explorer of fossil mammals and obscure documents, who has helped in many ways throughout the project.

The second group of people are less well-known to me personally, but I wish to record my gratitude to that indefatigable band of scholars who, over the course of several decades, have undertaken the heroic task of transcribing, annotating, and making available online the major part of Darwin's notebooks, diaries, letters and other manuscripts. I (like many others) have greatly benefitted from their brilliant work and open spirit in providing easy access to, and erudite commentary on, this inexhaustibly rich source of material.

Other colleagues have provided important help in various ways, providing access to specimens and documents, helping with identifications, answering queries, reading draft text, and providing pictures. It is a pleasure to thank Marina Aguirre, Patrick Armstrong, Paul Bahn, Marcelo Beccaceci, Fahema Begum, Mariana Brea, Adriana Candela, Fredy Carlini, Jessica Cundiff, Dianne Edwards, Marcos Ercoli, Howard Falcon-Lang, Marco Ferretti, Folkmar Hauff, Caroline Lam, Vanesa Litvak, Robert McAfee, Dimila Mothé, Geraldine O'Driscoll, Roula Pappa, Frances Parton, Paul Pearson, Dan Pemberton, Robbie Phillips, Roberto Portela-Miguez, Matt Riley, Adrian Rushton, Karolyn Shindler, Bruce Simpson, Peter Skelton, Tony Stuart, Charles Underwood, Diego Verzi, Colin Woodroffe, Frank Zachos, Marcelo Zárate and Alfredo Zurita. Special thanks to Mauricio Antón for his artwork, to Kevin Webb, Lucie Goodayle and Harry Taylor for fossil photography, and all the team in NHM Publishing for their support throughout.

ADRIAN LISTER

CHAPTER 1

The making of a naturalist

ON 25 OCTOBER 1831, a young man boarded a ship preparing for a voyage around the world. His name was Charles Darwin, and it was a journey that would change the way humanity thinks about itself and its relationship to the living world. Within months of his return, Darwin was writing his first, feverish notes on evolution that would lead, more than 20 years later, to *The Origin of Species*, one of the most influential books ever published. Later still he said: 'The voyage of the *Beagle* has been by far the most important event in my life and has determined my whole career.' What he saw on the five-year voyage led him to think deeply about the natural world, and to question received opinion about its origins. He also collected specimens – thousands of them – and these specimens, studied in the field and on his return to England, provided vital evidence in support of his ideas, especially on the theory of evolution.

Darwin's natural history interests on the *Beagle* voyage were all-embracing. He sieved minute organisms from the sea, he collected plants, insects, birds and mammals, but less widely appreciated was his passion for geology. By far the largest portion of his time – in South America and elsewhere – was devoted to observing and recording rocks and understanding the formation of the lands he was visiting. Together with the rocks came fossils, from countless shells to petrified trees and the remains of giant mammals. On the very first page of *The Origin of Species*, Darwin summarized the two key factors that had, thanks to his observations

Charles Darwin aged 31, four years after his return from the *Beagle* voyage.

on the *Beagle* voyage, led him to accept the theory of evolution. One was the distribution of animal and plant species, especially on islands. The second was the evidence of fossils – especially the relation of extinct forms with living species. While the former – particularly the biota of the Galápagos Islands – has received widespread attention, the latter is far less appreciated and forms the subject of this book.

A young man of promise, and extremely fond of geology

The circumnavigation by HMS *Beagle* in 1831–1836 has become famous for carrying Darwin on his voyage of discovery, but that was not its original purpose. The British Admiralty commissioned HMS *Beagle* to complete a survey of South American waters that she had begun on an earlier voyage in 1826–1830. The area was important to the British government for both trade and military purposes, and accurate charts of sea passages and harbours were essential. After surveying the coast of South America, the *Beagle* was to continue westwards, investigating coral reefs in the Pacific and Indian Oceans, and completing an accurate series of longitude measurements around the globe. For the latter purpose she carried an unprecedented 22 different chronometers representing the then state-of-the-art in nautical measurement.

The *Beagle*'s captain, Robert FitzRoy (1805–1865), had sailed with the ship on her earlier survey of South America, starting as lieutenant and later taking command. He was not only expert in naval skills of surveying, meteorology and oceanography, but considered himself a man of science. On the previous voyage he had regretted that 'so good an opportunity of ascertaining the rocks and earths of these regions should have been almost lost', so for the new expedition he proposed to Francis Beaufort, the Navy's hydrographer, that 'for collecting useful information during the voyage… some well-educated and scientific person should be sought'. He also, on a more personal level, wished for an intelligent companion with whom to discuss matters of science. Beaufort, in search of a suitable candidate, wrote to a contact at Cambridge University, the mathematician George Peacock, who wrote in turn to the Revd John Stevens Henslow (1796–1861), the Professor of Botany. Henslow first considered himself as a candidate, but was dissuaded by his wife; he then put the idea to Cambridge naturalist Revd Leonard Jenyns, who mulled it over for a day before deciding he could not leave his parish. The two of them then thought of Darwin, a student at the University training for holy orders but better known to

Robert FitzRoy, Captain of the *Beagle* and pioneering meteorologist. He sought a scientific companion for the voyage.

them as a keen naturalist. As later recounted by FitzRoy: 'He [Henslow] named Mr Charles Darwin, grandson of Dr Darwin the poet, as a young man of promising ability, extremely fond of geology, and indeed all branches of natural history.' Henslow wrote to Darwin with the news: 'I consider you the best qualified person I know of who is likely to undertake such a situation…The voyage is to last two years and if you take plenty of books with you, anything you please may be done.' Darwin received the fateful letter at his home in Shrewsbury on 29 August 1831. His father, Dr Robert Darwin, at first refused permission for the journey, until persuaded by Charles's maternal uncle, Josiah Wedgwood II, that the voyage would benefit Charles's character and would not damage his future career prospects as a clergyman.

Charles Darwin was as well-prepared for the *Beagle* voyage as any young naturalist could have been. Born on 12 February 1809 in Shrewsbury, the county town of Shropshire in the west of England, his mother died when he was eight, leaving his father, a prosperous country doctor and property-owner, to care for Charles, his brother and four sisters. By this time, Charles later recalled, he had already developed a 'taste for natural history, and more especially for collecting', amassing and organising pebbles, minerals and shells, as well as coins and other items. His elder brother, Erasmus, set up a laboratory in an outhouse, where the boys studied minerals and produced 'stinks and bangs', earning Charles the nickname 'Gas' at school.

Following the family tradition, Darwin went up to Edinburgh University in 1825, aged just 16, to study medicine. Medical studies were not to his liking, but he took the opportunity to attend courses in natural history, mineralogy and other subjects. Darwin professed to finding some of the lectures 'incredibly dull', but they introduced him to current geological debates, and he visited Edinburgh Museum to study minerals. An important influence on Darwin in Edinburgh at that time was Robert Grant (1793–1874), a lecturer in the medical school, who took him to the shores of the Firth of Forth to collect marine invertebrates, followed by their detailed examination, in which Darwin learnt both dissecting skills and how to use a microscope. Grant's particular interest was in primitive organisms then thought to represent a link between the plant and animal kingdoms, and Darwin made studies of seaweeds and bryozoans ('moss animals').

Darwin abandoned his medical studies in 1827 and went up to Christ's College, Cambridge the following year to study for holy orders, beginning with a so-called ordinary degree. There he greatly expanded his natural history interests. With like-minded friends he scoured the Cambridgeshire countryside in search of beetles. He attended lectures in botany given by Henslow, becoming part of the professor's inner circle and assisting him in collecting, preparing and mounting plant specimens.

In 1830 Darwin discovered the *Personal Narrative* by celebrated German explorer Alexander von Humboldt. Darwin was entranced by Humboldt's vivid descriptions of tropical scenery, and hatched a plan for an expedition to Tenerife in the Canary Islands, about whose natural beauty Humboldt had waxed eloquent. Darwin was also inspired by the fact that Humboldt had made significant scientific discoveries on his travels. He enlisted Henslow and some university friends, including Marmaduke Ramsay, to accompany him. In preparation for a trip in the summer of 1831, Darwin was 'crammed' in geology by Henslow, who had been Professor of Mineralogy until his resignation in favour of botany, and Adam Sedgwick (1785–1873), Professor of Geology. He continued his botanical studies and even took up Spanish. He wrote to his sister Caroline: 'My head is running around in the Tropics: in the morning I go and gaze at palm trees in the hot-house and come home and read Humboldt: my enthusiasm is so great that I cannot hardly sit still on my chair.'

Returning to Shrewsbury in mid-June 1831, Darwin acquired one of the basic tools of a field geologist, a simple instrument combining a compass with a clinometer for measuring the angle of slope of a rock. Learning to use it, he

RIGHT John Stevens Henslow, Darwin's Cambridge mentor, who recommended him for the *Beagle* voyage and took care of his specimens as they arrived back in England.

ABOVE Robert Edmond Grant, Edinburgh zoologist who took the 17-year-old Darwin collecting coastal organisms and taught him how to study them.

RIGHT Adam Sedgwick, Professor of Geology at Cambridge, whose crash course in field geology helped prepare Darwin for his work in South America.

'put all the tables in my bedroom at every conceivable angle and direction', and travelled through the Shropshire countryside mapping rocks. 'I am at present mad about geology', he wrote to a friend.

The trip to Tenerife, however, was not to be. Henslow excused himself through pressure of work; Darwin discovered that boats from England to the Canaries departed only from February to June; and worst of all, his friend Ramsay died on 31 July 1831. Yet within a few days, Darwin was packing his bags since, as part of his geological cramming, Sedgwick had offered to take him along on a two-week geological survey of the ancient rocks of North Wales. Sedgwick travelled to Darwin's home, and on 5 August they set out, travelling around 75 miles (120 km) across the mountains of Snowdonia to Anglesey in northwest Wales. During the trip, Darwin's theoretical knowledge of geology grew into field expertise. He learnt skills of observation, measurement and note-taking, and how to recognize and map different varieties of rock. Along the way, they visited caves where they found bones of ice-age animals, including the tooth of a woolly rhinoceros – startling evidence of the prehistoric world.

When the geologizing was over, Charles paid a visit to friends in Barmouth on the Welsh coast, then proceeded to the home of his maternal family at Maer in Staffordshire for a few days' shooting, before returning to Shrewsbury on 29 August to find letters from Henslow and Peacock waiting for him. The Tenerife trip might have been abandoned, but none of his preparations had been in vain, for they fortuitously stood him in very good stead for the much greater opportunities ahead. Instead of preparing to return to Cambridge for theological training, Darwin travelled to London to be interviewed by FitzRoy. FitzRoy felt – with good justification, as it turned out – that Darwin's passion for natural history, his good-humour and gentlemanly credentials, as well as his country skills of riding and shooting, would stand him in good stead for the voyage. Darwin began his hectic preparations, arriving in Devonport, where the *Beagle* was being readied, on 24 October 1831.

Setting sail

After a series of delays, the *Beagle* finally set sail on 27 December 1831. First stop was to have been Tenerife in the Canary Islands, which Darwin had longed to visit, but on arrival FitzRoy was informed that for fear of disease the *Beagle* had to fulfil 12 days' quarantine before landing. The Captain determined to press on,

The voyage of the *Beagle*, 1831–1836. The planned two years of the voyage expanded to five, three of them spent surveying the coasts of South America.

and Darwin was distraught: 'Oh misery, misery… we have left perhaps one of the most interesting places in the world…'. Within 10 days, however, the *Beagle* had crossed the Tropic of Cancer and landed at St Jago in the Cape Verde Islands. Here Darwin 'first saw the glory of tropical vegetation', and was overwhelmed by the beauty of trees, flowers and insects and their sheer abundance and diversity. He also recorded his first geological notes and collected his first fossils.

A month later they crossed the equator, and on 28 February 1832 landed at Bahia in Brazil, where Darwin was again enthralled by the tropical scenery. After some time at Rio de Janeiro, the *Beagle* arrived at Montevideo on the great river of La Plata in July 1832, and the serious work of the voyage began. The *Beagle* voyage can conveniently be divided into three phases: 27 months (1832–1834) on the Atlantic coast of South America; 15 months (1834–1835) on the Pacific coast; and a final 12 months (1835–1836) completing the circumnavigation and returning to England.

Life on board the *Beagle* was cramped. Only 90 ft (27 m) long, she accommodated a crew of 75 men, plus three natives of Tierra del Fuego whom FitzRoy had taken to England to be 'educated' and was now returning to their homeland. Darwin's accommodation was in the poop cabin, beneath the deck at the rear end of the ship. The room contained a large table on which the ship's

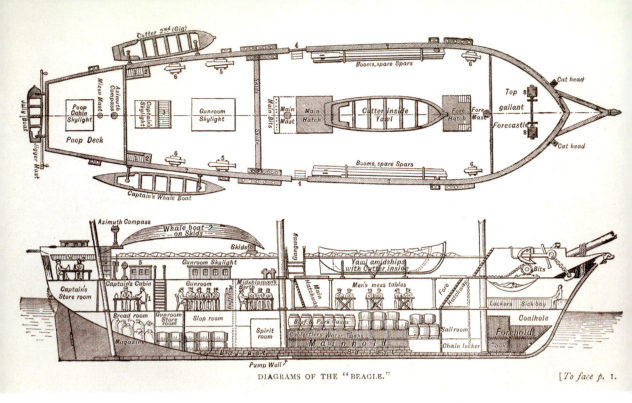

charts were laid out, and Darwin was allocated a chair and small space at the end of the table for working. He dressed and slept in the same room, his hammock slung above the table, and had a few small drawers for his clothes. Moreover, he shared this space with assistant surveyor John Lort Stokes and 14-year-old midshipman Philip Gidley. As he wrote to Henslow before departure, 'The corner of the cabin, which is my private property, is most woefully small. I have just room to turn around and that is all.' Conveniently, however, the *Beagle's* library, plus shelves for his own books, were in the same room, and for his specimens he was given a very small cabin at the opposite end of the ship. Conditions improved in June 1833 after FitzRoy purchased a second ship, the *Adventure*, to increase the surveying potential of the voyage, and delegated various officers to crew her, including Stokes. So for the following 18 months, until the *Adventure* was sold, Darwin had the cabin to himself (aside from its continuing use as a chart room). In the evenings, he habitually dined with the Captain in his cabin.

In FitzRoy's crew list, Darwin was listed as 'Naturalist', although his nickname on board soon became 'Philosopher'. The role of naturalist was by tradition fulfilled by the ship's surgeon, and the original incumbent of that position, Robert McCormick, felt himself usurped by Darwin's presence, leaving the ship only four months into the voyage, 'very much disappointed in my expectations of carrying out my natural history pursuits'. In general, however, Darwin's relations with his shipmates were extremely cordial. He and FitzRoy respected each other, and letters they exchanged when separated on the voyage reveal a kind of jovial friendship. Darwin soon learned, however, not to provoke FitzRoy's volatile temper; early in the voyage they argued about slavery (FitzRoy in favour, Darwin against) to the point where he thought he might have to leave the ship. But the storm passed, and Darwin could later write to FitzRoy: 'I think it far the most fortunate circumstance of my life that the chance afforded by your offer of taking a naturalist fell on me.' On the voyage they were

Reconstruction of Darwin in his cabin on board the *Beagle*, his hammock slung over the chart table, a skylight above. He frequently retired to his hammock when sea-sick.

scientific companions, just as FitzRoy had hoped. There was little or no conflict because FitzRoy was not at that point the religious fundamentalist that he later became, and any evolutionary inklings that Darwin may have had on the voyage (see Chapter 6) he kept strictly to himself. Other officers of the ship had scientific interests and aided Darwin in collecting and recording, sometimes at places Darwin did not visit because he was travelling elsewhere. His strongest friendship, which lasted the rest of his life, was with Lieutenant (later Admiral) Bartholomew Sulivan (1810-1890), who not only worked with Darwin on the *Beagle* voyage but sent him copious geological notes and some specimens from his later travels.

Darwin's most constant helper, however, was Syms Covington, who joined the ship aged 15 as 'Fiddler and boy to the poop cabin'. He became Darwin's *de facto* servant, a position formalized by FitzRoy in May 1833. Darwin only occasionally mentions him in his accounts of the voyage, but it is clear that Covington helped him on many occasions connected with fossil-hunting, for

Syms Covington joined the *Beagle* at 15 as ship's fiddler, but became Darwin's servant and helped him with collecting and recording his specimens, including many fossils.

example when specimens had to be extracted from intractable rocks. Covington remained Darwin's servant even after their return to England, until he emigrated to Australia in 1839.

The main inconvenience to the *Beagle*'s crew, occasioned by Darwin's presence, was the specimens he regularly carried onto the cramped vessel. First Lieutenant John Wickham, in particular, was 'always growling about me bringing more dirt on board than any ten men'. Fitzroy, half-jokingly, recalled '…our smiles at the cargoes of apparent rubbish which he frequently brought on board'. Fortunately for all concerned, Darwin regularly dispatched his specimens back to England.

The only known contemporary illustration of Darwin on board the *Beagle*. This humorous watercolour, believed to be by ship's artist Augustus Earle, shows Darwin (in top hat) explaining a specimen to a crew member while at his feet lie skulls and other bones, one labelled 'Tusk 4003 BC'.

The excursions begin

For two years the ship plied up and down the Atlantic coast of present-day Uruguay and Argentina, from the Plata in the north to Tierra del Fuego in the south, as the officers completed their surveys. Whenever the *Beagle* was in port, Darwin was off exploring the coast and the interior, on foot or on horseback, recording the geology and collecting fossils and other natural history specimens. On 22 September 1832 he first saw and collected the remains of fossil mammals. In the bay of Bahia Blanca he rowed ashore at a spot called Punta Alta, in the company of Captain FitzRoy and Lieutenant Sulivan, and in a low cliff found 'numerous shells and the bones of large animals'. This has been described by Richard Darwin Keynes, Charles's great-grandson, as a 'red letter day for biology', as it marked the beginning of the first line of evidence that eventually led him to question whether species had remained unchanged since their creation. At many places along the coast he also found abundant fossils of marine invertebrates, which contributed crucial evidence on changing past environments and the gradual uplift of the continent.

Darwin also made several major overland excursions, leaving the *Beagle* at one port and, by arrangement, meeting her again several weeks later further up the coast. For these expeditions he hired horses and local guides, but the routes were not without danger. The *Beagle* had arrived in South America at a time of unprecedented political turmoil, as recently independent former Spanish colonies waged border wars against each other and fought civil conflicts for internal control. On a ride from Maldonado in Uruguay in May 1833, for example, Darwin hired two men armed with pistols and a sabre, since the day before a traveller had been murdered on the road.

In August of the same year, with the *Beagle* anchored at the Rio Negro, Darwin set out on a series of overland rides in which he would cover some 1,200 miles (1,900 km) on horseback. The first leg took him 170 miles (270 km) north to Bahia Blanca where, arriving on 17 August, he went down the bay to Punta Alta and found more important mammalian fossils. The next and longest leg, 400 miles (640 km) north to Buenos Aires, was the most hazardous, and Darwin required

Map showing Darwin's main overland excursions in South America. Hiring local guides, horses and mules, he rode in total some 2,500 miles (4,000 km) in pursuit of geology, natural history and fossils.

PARAGUAY

CHILE

R. Paraguay

R. Paraná

BRAZIL

Copiapó

Huasco

ARGENTINA

R. Saldo

R. Paraná

Coquimbo (8)

R. Uruguay

Santa Fe

Illapel

Uspallata Pass

Bajada de Santa Fe

URUGUAY

Valparaíso (7) Mendoza

Rosario

Mercedes

Santiago

Piuquenes Pass

(3)

(4)

Buenos Aires

Maldonado

R. Plata

Montevideo

R. Salado

Concepción

(2)

Valdivia

R. Colorado Bahia Blanca

(1)

R. Negro

Patagones

St Joseph's (Valdes) Peninsula

Island of Chiloé

(6)

Chonos Archipelago

PATAGONIA

Port Desire

Port St Julian

(5)

Falkland Islands

R. Santa Cruz

Strait of Magellan

TIERRA DEL FUEGO

Beagle Channel

Cape Horn

clearance from General Rosas, the future dictator of Argentina, who controlled the region and was carrying out a ruthless campaign against the indigenous people. Rosas ordered a troop of soldiers to accompany Darwin and they paused every 30–50 miles (50–80 km) at the checkpoints manned by Rosas's militia. Darwin arrived in Buenos Aires on 20 September, and after only a week's pause he was off again, heading 300 miles (480 km) further inland up the Plata basin to the town of Santa Fe. There he was enchanted by luxuriant vegetation with 'beautiful flowers, around which humming birds were hovering', but fossil-hunting was never far from his mind. Darwin knew that huge bones had been discovered in this area 50 to 60 years earlier, including the shell of a giant armadillo-like creature and the celebrated skeleton of a giant sloth. He was not to be disappointed; he found key fossils, including the first evidence that horses had lived wild in the Americas before the arrival of humans (see Chapter 2).

Darwin returned to Buenos Aires by boat down the Paraná River, but a 'violent revolution' had broken out and the city was blockaded; 'I found to my utter astonishment I was a sort of prisoner.' Only by mentioning Rosas's name to a rebel general was he allowed into the city, accompanied by soldiers. He had been anxious to rejoin the *Beagle* ahead of her passage south to Patagonia, but on boarding the ship discovered that her departure was delayed while the officers completed their charts. He therefore took himself off on a final excursion of some 300 miles (480 km) across Uruguay, then known as Banda Oriental (i.e. the land to the east of the Uruguay River) where, at the home of a local landowner, he purchased arguably the most remarkable find of the voyage, the massive fossil skull of a bizarre and then unknown group of mammals (see Chapter 2).

The *Beagle*'s itinerary had led Darwin to some major discoveries, but they would not have happened without his energy and determination not to miss any opportunity. Fossils and geology were clearly paramount in these excursions; as he wrote to his cousin William Darwin Fox, while resting in Buenos Aires: 'My object in all this galloping was to understand the geology of those beds which so remarkably abound with the bones of large and extinct quadrupeds'. He also clearly revelled in the adventure, writing to his sister Caroline, 'I am become quite a gaucho, drink my mattee and smoke my cigar, and then lie down and sleep as comfortably with the heavens for a canopy as in a feather bed.'

Twice during her sojourn on the Atlantic coast of South America, the *Beagle* visited Tierra del Fuego at the southern tip of the continent, and then sailed some 400 miles (640 km) eastwards to the Falkland Islands, where she spent

A watercolour by the ship's second artist Conrad Martens of HMS *Beagle* in Tierra del Fuego. Some of Martens' pictures were commissioned by Darwin and hung in his house at Downe, UK.

a month in each of the southern summers of 1833 and 1834. Here Darwin collected fossils of immense importance, the oldest then known outside Europe.

There were natural hazards to be negotiated, too. On 2 October 1832, a party went ashore at Monte Hermoso in the bay of Bahia Blanca, Darwin among them. A raging storm blew up and the *Beagle* was unable to approach the shore. The party spent two nights on the beach with very little food before being rescued; Darwin recorded that they made a 'sort of tent' with their boat's sails, and 'managed pretty well till the rain began, and then we were sufficiently miserable'. FitzRoy commented: 'Mr Darwin was also on shore, having been searching for fossils, and he found this trial of hunger quite long enough to satisfy even his love of adventure.' In January 1833, at sea off Cape Horn, a terrible storm wrecked many ships and almost brought the *Beagle* voyage, and the lives of all on board, to an abrupt end. The ship weathered the storm, but FitzRoy noted that 'Mr Darwin's collections, in the poop and forecastle cabins on deck, were much injured.' Two weeks later, FitzRoy recorded that 'we passed into a large expanse of water, which I named Darwin Sound, after my messmate, who so willingly encountered the discomfort and risk of a long cruise in a small loaded boat'.

Andean adventure

On 10 June 1834 the second major phase of the voyage began, as the *Beagle* passed through the Strait of Magellan and entered the Pacific Ocean. For the next 15 months she surveyed the coast of Chile (with a brief sojourn in Peru at the end), and Darwin once again spent as much time as possible on shore. At first, for around three months in total, the ship explored the islands of Chiloé and the Chonos archipelago, where Darwin studied shell-beds, saw massive trunks of petrified wood, and found evidence that the western as well as the eastern side of the continent had recently been uplifted.

In August and September 1834 the first of Darwin's overland excursions in Chile took him on horseback, with two guides, from Valparaíso to Santiago and back, a round-trip of six weeks, collecting marine fossils en route. The magnificent chain of the Andes, always in sight, dominated Darwin's geological and palaeontological work through this phase of the voyage. He and the *Beagle* crew also chanced to witness volcanic eruptions and a major earthquake – dramatic first-hand experiences that contributed to his thinking about mountain-building and the uplift and subsidence of continents and oceans (see Chapter 4).

In March 1835 Darwin set out on his most ambitious journey of the voyage – to cross the Andes. He was accompanied by a Chilean guide, Mariano Gonzales, plus a mule-driver, 10 mules (four for riding and six laden with food), and a mare with a bell round her neck for the mules to follow. From Santiago the party climbed some 13,000 ft (4,000 m) to the Piuquenes Pass, the watershed between the Pacific and Atlantic Oceans. Here Darwin was overjoyed to find fossils of animals that had once lived on the sea floor (see Chapter 4). After descending to the city of Mendoza on the other side, the party returned by a different pass, the Uspallata. This led to probably the most serendipitous of all Darwin's discoveries, as his path chanced to pass directly by a unique grove of upright, petrified trees (see Chapter 3). Soon after, he set off with another guide on his last major excursion of the voyage. They rode some 420 miles (675 km) northwards from Valparaíso, with the mountains to their right, exploring successive valleys as they passed them. By arrangement, the *Beagle* picked Darwin up at the town of Copiapó two months later, in July 1835.

The Cordillera (Andes) seen from Santiago, Chile, Darwin's starting-point for crossing the range. He ascended to 13,000 ft (4,000 m), where fossils of ancient marine creatures awaited him.

The British connection

In all his travels in South America, Darwin had been greatly assisted by an extensive network of British diplomats, landowners and merchants. It was thanks to the British chargé d'affaires in Buenos Aires that Darwin had gained his audience with General Rosas that allowed him to travel overland in relative safety. Another key contact was Edward Lumb, a prosperous English merchant; he and his wife accommodated Darwin at their Buenos Aires home, where he 'enjoyed all the comforts of an English country house'. They helped plan his overland trips and provided an introduction to Mr Keen, 'a very hospitable Englishman' living in Uruguay, leading to some of his most significant fossil finds. In Chile, Darwin spent much time at the home of Richard Corfield, an English merchant in Valparaíso whom Darwin had known at Shrewsbury School. Later, crucial assistance was provided by Alexander Caldcleugh, a British trader and plant collector in Santiago, who 'most kindly assisted me in making all the little preparations for crossing the Cordilleras' and drew him a detailed route map.

A map of the Andean foothills prepared for Darwin by Alexander Caldcleugh, one of a network of British expatriates who assisted Darwin in his travels.

Onward around the world

As the *Beagle* left South American shores in September 1835, the third phase of the voyage began. Sailing across three oceans gave Darwin the opportunity to ponder all that he had seen, while Covington helped him to bring some order to his notes, including the compilation of specimen lists. The monotony of the ocean was broken by a series of stops at volcanic and coral islands, including Tahiti in the Pacific, and the Cocos (Keeling) Islands and Mauritius in the Indian Ocean. Darwin's observations here led to major insights into the nature and formation of coral reefs (See Chapter 5). Significant fossil discoveries also awaited him around Sydney, Australia and especially in Tasmania. Despite the increasing lure of home, every stop provided an opportunity for addressing a new question, as on the island of St Helena in the Atlantic, where observations on living and fossil shells led him to deduce the age of the island and the history of its fauna (see Chapter 4). The potentially idle youth feared by his father now embodied his own declaration that 'a man who dares to waste one hour of time has not discovered the value of life'. Touching the coast of Brazil, he experienced again the tropical luxuriance that had captivated him nearly five years previously; and within two months they were home.

Keeping in touch

Darwin was not immune from homesickness during the voyage, and letters from his family, full of gossip as well as reports on the reception of his specimens, were a lifeline. Communications were frustratingly slow, however. As the *Beagle* moved from port to port, Darwin dispatched letters to his friends and family advising them where they should write to him over the coming months. For example, from Valparaíso on 10 March 1835 he wrote to his sister Caroline: 'After receiving this you must direct till the middle of November to Sydney, then till the middle of June to the Cape of Good Hope' – instructions for more than a year ahead. But if the *Beagle* missed any mail, the letters chased the ship round the globe. The Darwin sisters took turns to write monthly, but their letters could take up to seven months to arrive; one from his cousin William Darwin Fox took a whole year.

More stressful was the fate of the packages of specimens he sent home. Before the start of the voyage, Darwin's mentor John Stevens Henslow had agreed to receive and store his collections in Cambridge. From his letters to Henslow during the voyage, around nine different shipments can be identified between August 1832 and April 1835, all except two of them including fossils. The boxes, casks and barrels were presumably made by the *Beagle*'s carpenter, Jonathan May, and her cooper, James Lester. They were sent by ship mostly from Buenos Aires or Montevideo, the last shipments departing from Valparaíso and Coquimbo, Chile. Subsequent collections, from Australia and islands visited on the return leg, were relatively modest and were retained on the *Beagle* until her return.

The complex business of getting the collections home is exemplified by the specimen Darwin considered to be his most valuable – the bizarre skull of an extinct giant mammal (identified as *Megatherium* by Darwin but later named *Toxodon*) that he purchased in November 1833 at a farm in a remote part of western Uruguay (see Chapter 2). The farm was on a stream flowing ultimately into the Uruguay River and the Plata estuary. As Darwin set out to ride the 150 miles (240 km) back to Montevideo, he left the precious skull in the care of Mr Keen with whom he had been staying. Keen packed it up and sent it down river, addressed to his friend Edward Lumb in Buenos Aires, who was to forward it to

England. The following March, from the Falkland Islands, Darwin wrote to Lumb: 'I am very anxious that the Megatherium head, which Mr Keen procured for me, should not be lost.' In May Lumb wrote to reassure him that he had received the skull, and that it had been dispatched on a commercial vessel, the *Bassenthwaite*, bound for Liverpool. At the same time he wrote to Henslow to alert him to the arrival of the package. Henslow informed Darwin's father, who would be paying for the carriage from Liverpool, and requested that the package be sent direct to the Royal College of Surgeons in London, where the curator, William Clift (1775–1849), would receive it. A friend of Lumb's was travelling on the same ship and saw the consignment though British customs. It duly arrived at the College in August 1834, but Darwin was not informed of its safe arrival until his sister Caroline wrote to him in December that their brother Erasmus, who was the family's contact in London, 'says a box with bones came to England by Liverpool in August, and he thinks from Buenos Aires'. Darwin received this news only in August 1835 in Lima, 21 months after entrusting the skull to Mr Keen.

A view to posterity

Darwin was a prolific writer throughout the *Beagle* voyage. He always carried a pocket notebook for recording immediate observations in the field – 15 notebooks were filled during the five years of the voyage. At leisure on the ship or during excursions he wrote up more detailed thoughts and observations – principally the 1,383 pages devoted to geology (including fossils), now known as the *Geological Diary*, and 368 pages of zoological notes. He also kept a personal diary of his experiences that eventually ran to 779 pages. All of these writings, as well as his letters to Henslow and his family, contain vital information concerning his fossil-collecting.

Henslow's decision to send the mammalian fossils to William Clift before Darwin's return was an exception; other parts of the collection were retained in Cambridge. The bones had been shown to specialists and caused quite a stir, and Clift had undertaken to clean them of sediment and repair them where necessary. When Darwin received this news from Henslow, he was 'in great fear lest Mr Clift should remove the [specimen] numbers', writing anxiously to both Henslow and his sister Caroline, the latter so that she might ask Erasmus to call on Clift to emphasize the point. This he duly did, and Clift reassured the family that he had 'observed that many of [the specimens] had numbers affixed to them, and these I took special care not to displace'.

THE SPECIMENS

For Darwin his specimens were of prime importance. Even before the start of the voyage, he knew that precise recording of the location where specimens were found, and in the case of rocks and fossils the stratum in which they lay, would be crucial to the scientific interpretation of his collections. So, before the *Beagle* voyage, Darwin prepared numbered labels which were then stuck to each specimen (or its container) on collection, and its provenance recorded by the same number in his notes – in the case of fossils the *Geological Diary*. Each label bore a number from 1 to 999 but there were four sets in different colours, each indicating a prefixed number: white = no prefix (1–999), red = 1 (1,000–1,999), green = 2 (2,000–2,999) and yellow = 3 (3,000–3,999). The idea behind this, Darwin explained, was that 'when unpacking, a single glance tells the approximate number'. The labels were all-important; on sending his first consignment of fossil bones to Henslow in November 1832, he stressed that should the boxes be unpacked, '*care must be taken… not to confuse the tallies*'; and later, 'All the interest which I individually feel about these fossils, is their connection with the geology of the Pampas, and this entirely rests on the safety of the numbers.'

A specimen of fossil wood bearing Darwin's original coloured label. The green label 705 signifies no. 2705, collected in 1835 from the Andes of Mendoza. The yellow oval label shows the current Natural History Museum registration number.

At the same time, it was made clear to Clift, on Darwin's instruction, that sending the bones to the College for repair and study did not necessarily imply that they would be deposited there permanently. Even before the start of the voyage, Darwin was exercised about the final destination of his anticipated collections. He assumed that he would be 'honour bound to give largely to the British Museum', as a result of being on a ship of the British Navy, but was not happy at the prospect, apparently because of the museum's poor record of curating or describing recently deposited collections. In the end, since all his

BELOW William Clift, curator at
the Royal College of Surgeons in
London, who received and prepared
Darwin's fossil mammals even while
the *Beagle* was still at sea.

ABOVE Richard Owen, Hunterian
Professor at the Royal College of
Surgeons, described and named
Darwin's fossil mammals, almost all
of them species new to science.

costs bar his accommodation on the *Beagle* had been borne by his father, Darwin's collections were considered as belonging to him, not to the Crown, so he could dispose of them as he wished. On his return, he donated all the fossil bones to the Hunterian Museum of the Royal College of Surgeons, on condition that they prepare sets of casts to be donated to the British Museum, the Geological Society, the Universities of Oxford and Cambridge, and himself. Richard Owen (1804–1892), Clift's former assistant and recently appointed Hunterian Professor, declared that the fossils were 'highly desirable for the collection', and the College's board recommended that Darwin's offer be accepted on the conditions proposed. During the period 1837–1840 Owen published names and descriptions of Darwin's fossil mammals (see Chapter 2), but when he moved to the British Museum in 1856, the fossils remained at the Royal College of Surgeons. On the night of 10 May 1941 the College was badly damaged by bombing and many of its collections destroyed, but Darwin's fossils were among those that survived. In 1946 the College donated his collection to the British Museum (Natural History) – now the Natural History Museum, London – a somewhat ironic outcome given Darwin's original concerns.

Other groups of fossils were passed by Darwin to an array of specialists, in the expectation that each would publish an account of their findings – a very modern approach. Many did, some did not – a very modern outcome. Darwin took great care, and advice where needed, in selecting his specialists. As far as the fossils were concerned, molluscs (of which there were more species than in any other group) were divided among three experts. The first, George Brettingham Sowerby I, was one of four generations of Sowerbys who were expert amateur malacologists. The second, Edward Forbes, was curator at the Geological Society in London and a brilliant polymath. The third, Alcide d'Orbigny, a French naturalist, had undertaken his own expedition to South America in 1826–1833, thus overlapping with the *Beagle*'s voyage. Although they never met, Darwin became aware of d'Orbigny's presence a few months into the voyage, writing to Henslow: 'By ill luck the French Government has sent one of its collectors to the Rio Negro where he has been working for the last six months, and is now gone round the Horn. So that I am very selfishly afraid he will get the cream of all the good things, before me.'

After Darwin's return, however, he and d'Orbigny established cordial relations and Darwin sent him quantities of fossil molluscs for identification, d'Orbigny equating many of them with species he had erected from his own collections. Darwin later wrote that 'considering that I had no claim on M. d'Orbigny's time, I cannot express too strongly my sense of his extreme kindness'. Nonetheless, Darwin at times sent the same samples

The French naturalist Alcide d'Orbigny, whose collecting tour of South America partly coincided with Darwin's. A specialist of fossil shells, he later identified many of Darwin's specimens.

to several of his specialists for alternative views, and it is clear there was a degree of rivalry among them. In a letter of February 1845, d'Orbigny writes: '… as I expected, your magnificent collection of fossils shows no contradiction with those that I saw in America. The apparent differences were entirely owing to the incorrect determinations of Mr Sowerby'. Sowerby in turn criticized some of the identifications of d'Orbigny, and Darwin at times had to make judicial choices. Other groups of fossils – plants, brachiopods, sea urchins, bryozoans – were also sent to relevant specialists (see Chapters 3 and 4), and most of these collections eventually found their way to the Natural History Museum, London. More important specimens were presented, as part of Darwin's extensive rock collection, to the Sedgwick Museum of Geology in Cambridge, and a few items have found their way to other museums around the world.

Darwin and evolution before the *Beagle*

Unlike Alfred Russel Wallace (1823–1913), whose expeditions to the Amazon and to Southeast Asia were partly undertaken 'with a view to the theory of the origin of species' (and who discovered the principle of natural selection independently of Darwin), evolutionary questions were not uppermost in Darwin's mind as he embarked on the *Beagle* voyage.

There can be little doubt, however, that Darwin was fully aware of the debates over 'transmutation' taking place around him. His own grandfather, Erasmus Darwin (1731–1802), had expounded evolutionary ideas in poetry and prose. Most widely known were the ideas of Jean Baptiste Lamarck (1744–1829), the French naturalist who proposed that organisms have a natural tendency to progress from simple to complex, and that this occurred due to the way different parts of the body were used or not used in relation to the environment. In Edinburgh, Robert Jameson, whose lectures Darwin had attended, published an anonymous article in 1827 in which he posited replacement of extinct species with those still living. Robert Grant, too, was a fervent adherent of evolutionary views; Darwin later recounted that on one of their collecting expeditions Grant had suddenly 'burst forth in high admiration of Lamarck'. Darwin later stated that neither listening to Grant, nor reading Lamarck or Erasmus Darwin, had had any effect on his mind, a claim which his biographer Janet Browne suggests was somewhat disingenuous, since at the very least he must have absorbed the concept of evolution and the contemporary debates surrounding it.

By far the most important influence on Darwin during the *Beagle* voyage was the newly-published treatise *Principles of Geology* by Charles Lyell (1797–1875). The first of its three volumes was presented to Darwin by FitzRoy before their departure, and he received the others during the course of the voyage. The underlying principles of Lyell's geology were, first, that anything evidenced in the rock record could be explained by processes that can be seen operating in the present day (uniformitarianism); and second, that major changes (such as the uplift of mountains) can arise as the cumulative effect of numerous small increments. These ideas were in stark opposition to the 'catastrophist' view, promulgated by the celebrated French anatomist Georges Cuvier (1769–1832), whereby major Earth changes were the result of occasional cataclysmic upheavals. Darwin's teachers, the Reverends Henslow and Sedgwick, both subscribed to this view, and to the idea that the more superficial geological sediments were the result of Noah's flood, seen as the most recent of the 'catastrophes', although Sedgwick was to renounce this idea in 1830 and Henslow in 1836. Darwin, under their influence, began the voyage as a catastrophist, but was soon persuaded by Lyell's views, enthusing about 'the wonderful superiority of Lyell's manner of treating geology, compared with that of any other author'. Lyell's philosophy of incremental change through natural processes became not only the cornerstone of Darwin's geological work, but ultimately, transferred to the organic realm, it provided the intellectual framework for his theory of evolution.

GEOLOGICAL TIME

The modern subdivision of geological time was incomplete at the start of the *Beagle* voyage in 1831, but was to make major advances during the 1830s and 1840s. The fossils Darwin collected span a vast range of geological time, from some 400 million to only a few thousand years ago, and his terminology for their likely ages reflects the period in which he wrote. Since the mid-18th century, geological strata had been divided into 'Primary', 'Secondary' and 'Tertiary' divisions. The 'Primary' was the oldest, and was defined partly on lacking any observable fossils, although there was confusion over whether it referred to a type of rock or the period of time in which it formed. The term was dropped, following the recommendation of Sir Charles Lyell in 1833, but the 'Secondary' and 'Tertiary' divisions remained current and were the terms habitually used by Darwin in his geological writings during the voyage. He also occasionally employed the term 'Transition', which had been introduced to refer to rocks intermediate between 'Primary' and 'Secondary' and which contained the oldest known fossils.

In 1841, as Darwin was preparing his voluminous geological observations for publication, John Phillips formerly named the Palaeozoic, Mesozoic and Cenozoic Eras, roughly corresponding to the old 'Primary/Transition', 'Secondary' and 'Tertiary' periods. Darwin employed the term 'Palaeozoic' in his *Geological Observations on South America*, published in 1846, although he retained 'Secondary' and 'Tertiary' for the two later periods, and we employ the latter term (in formal use until 2004) when describing Darwin's observations.

The Paleozoic is now dated to 541–252 million years ago, and among its characteristic fossils are trilobites, brachiopods and early fishes; the Mesozoic ('Secondary') spans 252–66 million years ago and is best known for dinosaurs and ammonites, while the Cenozoic spans from 66 million years ago to the present day, and is often known as the 'Age of Mammals'. The Cenozoic incorporates the old Tertiary division up to around 2.6 million years ago and the Quaternary (the time of the most recent ice ages) thereafter. The latter term was added in 1829 but was not employed by Darwin and did not enter common usage until much later.

ERA		PERIOD OR EPOCH	AGE (millions of years)
CENOZOIC	QUAT-ERNARY	Holocene	0.012
		Pleistocene	2.6
	TERTIARY	Pliocene	5.3
		Miocene	23
		Oligocene	34
		Eocene	56
		Paleocene	66
MESOZOIC	SECONDARY	Cretaceous	145
		Jurassic	201
		Triassic	252
PALEOZOIC	PRIMARY / TRANSITION	Permian	299
		Carboniferous	359
		Devonian	419
		Silurian	443
		Ordovician	485
		Cambrian	541

1. Period names shown vertically were recognized in Darwin's time but are no longer in use, except for the Quaternary.

2. The vertical (time) axis is not to scale.

3. The Precambrian is not shown.

4. Epochs rather than periods are specified for the Cenozoic.

CHAPTER 2

~

Giant mammals

THE SKULLS AND BONES OF EXTINCT mammals were the crowning glory of Darwin's fossil collecting in South America, not only for him but for the eager recipients of the cargoes he sent home. They were the discoveries that made his name known beyond his immediate circle; when the first consignment of fossil bones arrived at the Royal College of Surgeons in 1833, the puzzled curator, William Clift, recorded them as apparently 'from a Mr Darwin at Rio de la Plata'. Only a few months later, however, after the specimens had been exhibited in Cambridge, his friend Frederick William Hope wrote to tell him that his 'name was in every mouth'. Even more significantly, Darwin later credited the fossil mammals as one of the two main factors that led him to embrace the reality of evolution.

The effect on the young naturalist of finding and excavating the huge skulls and bones of evidently extinct giant mammals, soon after arriving at the coast of Argentina, can hardly be exaggerated. He wrote to his sister Caroline how his former pastimes paled into insignificance: 'The pleasure of the first day's partridge shooting or first day's hunting cannot be compared to finding a fine group of fossil bones, which tell their story of former times with almost a living tongue.' Even when the *Beagle* had moved round to the Pacific coast of South America, and Darwin was exploring older rocks with fossils of marine organisms, he wrote excitedly to Henslow: 'I have just got scent of some fossil bones of a Mammoth, what they may be, I do not know, but if gold or galloping will get them, they shall be mine.'

The giant sloth, *Megatherium americanum*, feeding on a high tree, its vertical posture supported on the tripod of its hind legs and tail.

It is remarkable that at his main hunting-ground at Punta Alta, in an area of some 500 x 500 ft (150 x 150 m), the relatively small number of fossils that Darwin collected should have demonstrated no fewer than seven distinct types (genera) of mammal, while six more were collected from other sites. Of these 13, only two were known at the time, and six were named on the basis of Darwin's specimens. Many of the species discovered by Darwin are now celebrated elements of the South American fossil fauna of the last ice age, some 100,000 to 12,000 years ago.

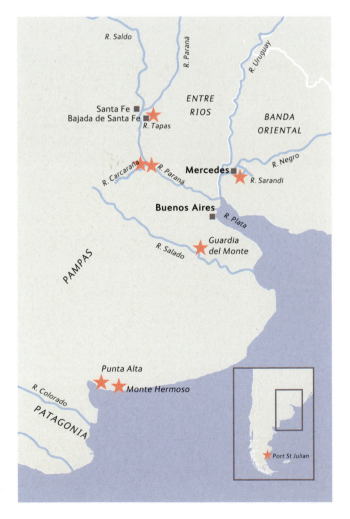

The localities where Darwin found fossil mammals, all of them in present-day Argentina and Uruguay.

The great beast

On only the second day of his fossil-hunting in South America, Darwin found the largest and heaviest single fossil of the entire voyage, belonging to the largest and heaviest land mammal ever to live in South America. On 23 September 1832, while a party from the ship was sent to catch fish, Darwin walked to Punta Alta, 'and to my great joy found the head of some large animal imbedded in soft rock'. He relates that it took him nearly three hours to extract the skull from the cliff, and he did not get it on board until several hours after dark.

RIGHT The partial *Megatherium* skull, some 20 in (50 cm) long, discovered by Darwin. A vertical slice was cut from the skull by Richard Owen in the 1830s to expose the structure of the molars. The slice (left) was rediscovered at Darwin's home, Down House, in 2017.

RIGHT The skull fragment seen from underneath. Its eroded surface and the hollow pulp cavities of the molars are probably the side exposed in the cliff at Punta Alta first spotted by Darwin.

The general form of the fossil was immediately evident – Darwin described it as 'the upper jaw and head of some very large animal, with four square hollow molars, and the head greatly produced in front'. When complete it would have been around 3 ft (1 m) in length, and Darwin speculated that it belonged to *Megatherium*, the 'great beast' as it had been named by celebrated French anatomist Georges Cuvier. This was the only giant South American fossil mammal known to science at the time.

Darwin identified several of his later finds as *Megatherium* too, but while most of them were subsequently shown to be different genera, his hunch on the first skull was correct. Ironically, he was strengthened in his opinion by finding pieces of giant bony plates in the same deposit; as will be discussed later, these were at the time thought to be the armour of *Megatherium* but were soon shown to belong to a different animal altogether.

The huge skull was not the only specimen of *Megatherium* that Darwin found at Punta Alta. He also recovered part of the left side of another skull, and the back portion of a third. He did not record when these were discovered, but the latter specimen, at least, is very likely to have been found during the first season at Punta Alta – September to October 1832 – rather than the second nearly a year later, since by the summer of 1833 the specimens were clearly in England and had caused quite a stir.

The earliest discovery of *Megatherium* had been a largely complete skeleton unearthed on the banks of the Luján River, west of Buenos Aires, in 1788. Discovered by a Dominican friar, Manuel Torres, it was sent to Madrid, where it became the first fossil mammal skeleton ever to be assembled and mounted for public display. An accurate drawing of the skeleton was sent to Cuvier, and although he never saw the actual bones, he described it in some detail and in 1796 named it *Megatherium americanum*. The skeleton was some 13 ft (4 m) long, with massive hindquarters, powerful front limbs, and huge, curved claws at the ends of its toes. Cuvier placed it within his 'edentate' order of mammals, along with the living sloths, anteaters, armadillos and some others, and presciently suggested that it was a gigantic sloth.

The living sloths are tree-dwelling, leaf-eating mammals exclusive to the forests of South and Central America. Slow-moving, they spend much of their time suspended upside-down from the branches of trees. They are relatively small

A living three-toed sloth, *Bradypus variegatus*, from the forests of South and Central America. The relationship between the living and fossil sloths was, for Darwin, a striking example of his law of succession of types.

mammals, their body weight rarely exceeding 11 lb (5 kg). The recognition that the giant *M. americanum* – now estimated to have been a thousand times heavier – belonged to the same group, was a triumph of comparative anatomy, and the contrast between this massive, ground-dwelling creature and its living relatives added to the astonishment it evoked in scientists and the public alike.

Darwin had an English translation of Cuvier's treatise with him on the *Beagle* and was familiar with *Megatherium*. However, a more immediate circumstance was to make Darwin's finds especially timely when they were unpacked in Cambridge. Shortly before Darwin's arrival in South America, huge bones had been discovered at the Salado River south of Buenos Aires. They came to the attention of the British Chargé d'Affaires, Woodbine Parish, who was interested in scientific matters, geology in particular. He sent Mr Oakley, a carpenter in his employ, to search the site for more bones. The result was an almost complete skeleton of *M. americanum*, which Parish brought back to England on his return in 1832, and exhibited at the Geological Society of London. He then donated the bones to the Royal College of Surgeons, whose collection of comparative anatomy was the finest in the country. William Clift, the curator of the collection, bestowed a great deal of time and care on the precious skeleton, only the second known after the Madrid specimen, with the intention of mounting it in the College's Hunterian Museum. Certain parts were missing, however.

Darwin's first consignment of bones was dispatched from Montevideo in late November 1832, and arrived in England the following January. In June, at the Cambridge meeting of the British Association for the Advancement of Science, Darwin's fossil finds were displayed by William Buckland, doyen of British palaeontology. Clift, fresh from his work on Parish's skeleton, examined the skull and recognized that the back portion recovered by Darwin was precisely the part missing from Parish's specimen. As Darwin's friend Frederick William Hope wrote to him: 'from sending home the much desired bones of Megatherium your name is likely to be immortalised. At the Cambridge meeting of naturalists your name was in every mouth & Buckland applauded you as you deserved'. His family, too, were reassured about the value of the voyage, his sister Caroline enthusing: 'I give you joy my dear Charles on having found these bones that delight the learned so much.' Even FitzRoy caught the spirit, writing to Darwin during one of the latter's land excursions in August 1833: 'Your home (upon the waters) will remain at anchor near the Montem Megatherium until you return' – a whimsical reference to Punta Alta where Darwin had found the *Megatherium* remains.

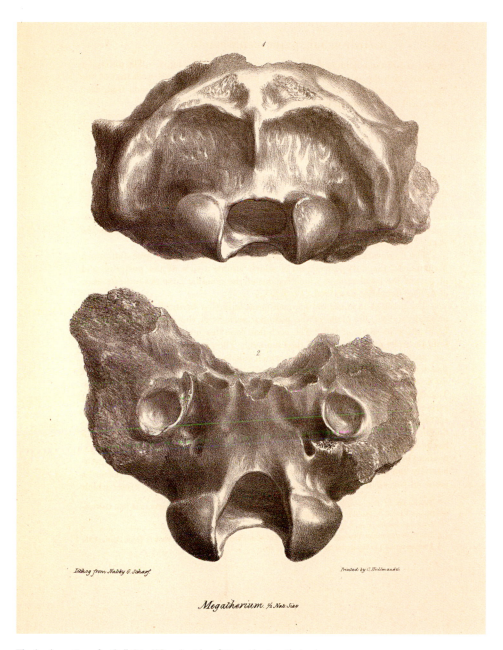

Lithog from Nakby G. Scharf

Printed by C. Hullmandel.

Megatherium. ⅓ Nat. Size

The back portion of a skull, 8 in (20 cm) wide, of *Megatherium* that
put Darwin's name 'in every mouth'. The illustration, from Owen's 1840
monograph, shows the skull from behind and underneath.

Darwin had read about Parish's skeleton in the newspapers as early as November 1832. He sought out Mr Oakley in Buenos Aires, and established to his satisfaction that the *Megatherium* skeleton had been found in the same strata as his own finds. He declared himself 'quite astonished that such miserable fragments of the Megatherium' should have attracted such attention, but was later 'much pleased' when he learnt how his finds had completed the jigsaw of this celebrated creature.

The intense public as well as scientific interest in the *Megatherium* finds can be understood in terms of their historical context. In the decades before the discovery and public awareness of dinosaurs, giant ice-age mammals such as the Irish elk, the mammoth and the mastodon were the poster-children of prehistoric

Megatherium as it appears in Owen's monograph of 1851 – correctly shown walking on the outer edges of its feet, but still in a horizontal posture.

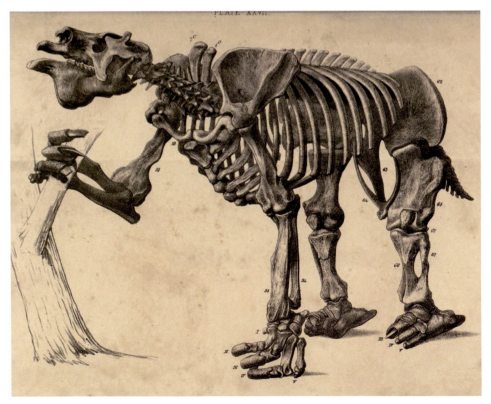

monsters, and none more so than *Megatherium*. The discoveries were the subject of newspaper articles, posters and public lectures, while in contemporary literature the animal became a metaphor for a lazy, ponderous or dysfunctional person or thing. As Darwin noted in his diary, even in South America the finds had become a source of legend. Local people told him that there existed an animal larger than an ox, with great claws and a snout, to which was given the name 'gran bestia'. Darwin surmised that the legend might have arisen from the finding of preserved skeletons of *Megatherium*.

In the scientific community, the Parish skeleton led both Clift and Buckland to write papers on *Megatherium*, but it was Richard Owen who reported in great detail, first on Darwin's specimens, and later in an exhaustive 130-page monograph covering all known finds of the species. Cuvier had set the scene with his description of the Madrid skeleton, considering its body ill-proportioned and its parts incongruent – a highly unusual conclusion from the man famous for showing that every species was designed as a harmonious whole. Buckland, in defence of the Creator, suggested that the powerful, clawed limbs had been used for uprooting plants, while the heavy backside had enabled the animal to stand on three limbs, freeing one forelimb for feeding on trees. This idea was at first accepted by Owen, but he later saw that the massive tail and hind limbs could form a tripod enabling the animal to stand almost upright, towering 13 ft (4 m) into the air to wrest branches from trees. This image has formed the basis for most paintings, models and mounted skeletons ever since (see p.34). It was immortalized in stone when, in 1854, the celebrated outdoor display of prehistoric animals was opened at Crystal Palace in south London. The sculptors were closely advised by Richard Owen, and *Megatherium* was – and remains – one of the stars of the show.

Since Owen's time, much has been added to our understanding of this legendary creature. *Megatherium americanum* is now estimated to have weighed between 4 and 6 tonnes – equivalent to a bull elephant. Remarkable evidence of its locomotion was discovered at Pehuén-Có, Argentina, in 1986 when tidal action uncovered a surface of hardened clay covered with animal footprints believed to date from around 14,000 years ago. Many species are represented, but the most striking feature is a row of 35 large footprints, the size and shape of which can only fit *M. americanum*. They suggest slow, about 3–6 ft (1–2 m) per second, largely bipedal locomotion, with the weight periodically dropping down onto all fours. As with other giant sloths, the feet were turned so that most of the animal's

Preserved footprints of *Megatherium*, each some 35 in (90 cm) in diameter, on the coast at Pehuén-Có, Argentina. The site is close to where Darwin collected fossils of the species, but was discovered only in the 1980s.

weight was borne on their outer edges. Coincidentally, the footprint site is on the bay of Bahia Blanca, only 25 miles (40 km) east of Punta Alta where Darwin collected fossils of *Megatherium* some 150 years earlier.

As for diet, a recent, detailed study of the skull and teeth indicates a narrow snout and prehensile lip typical of an animal that selects individual leaves, fruits or twigs. Because the teeth bear sharp ridges suggestive of cutting rather than grinding action, some researchers suggested that *Megatherium* might have incorporated some meat in its diet. However, this wear-pattern can also be related to soft-leaf browsing, and recent studies of carbon and nitrogen isotopes in the fossil bones confirm a purely vegetarian diet.

Megatherium americanum was not the only giant sloth species among Darwin's collection; three more were identified by Owen, all of them new to science. The

The extinct, 1.5-tonne ground sloth, *Mylodon darwinii*, named by Richard Owen from fossils collected by Darwin in South America.

first was a species that Owen named *Mylodon darwinii* after its discoverer. It was not recorded by Darwin at the time of its discovery, although he later indicated it was from Punta Alta, and it was probably the specimen referred to in his letter to Henslow in October 1832, where he related having found 'the lower jaw of some animal, which from the molar teeth, I should think belonged to the Edentata'. A year later he referred to it as 'the animal of which I sent the jaw with 4 small teeth' and appended a sketch that leaves no doubt he was referring to the mandible that is now one of the treasures of the surviving Darwin collection.

The finding of a second lower jaw was described in more detail; Darwin wrote in his diary that on 8 October 1832, after breakfast, he walked to Punta Alta and 'obtained a jaw bone which contained a tooth'. His immediate thought was of *Megatherium*, but he later more cautiously suggested only that it was one of the Edentata. Owen, when he came to describe it in 1840, assigned it to the only other extinct giant sloth that had been named prior to Darwin's discoveries. This was *Megalonyx*, a fossil then almost as famous as *Megatherium*, especially in the USA. Its remains had first been unearthed in 1796 in what is now West Virginia, and came to the attention of polymath (and later President) Thomas Jefferson. From its massive claws he supposed it to be a carnivorous animal related to the lion, and suggested the name 'Great-claw or Megalonyx'. Subsequently the animal was recognized to be a giant sloth, and for a while the difference between it and *Megatherium* was unclear, until Cuvier showed them to be distinct. *Megalonyx* is

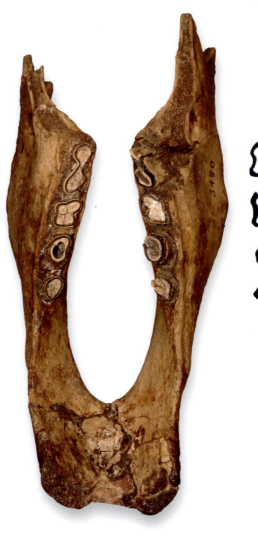

The lower jaw of *Mylodon darwinii* from Punta Alta, 14 in (35 cm) long. Alongside is Darwin's sketch of the teeth, included in a letter home. Such details are of great value in relating existing specimens to Darwin's original accounts.

now known to have been restricted to North America, but it was a reasonable conjecture on Owen's part that its remains might be found in South America too. He noted that the jaw, as well as retaining only a single, damaged tooth, was embedded in the lump of cemented gravel that Darwin had excavated, so that only its top edge was visible, and the bone, moreover, was crumbling. The specimen no longer exists – the latest record of its existence is in a catalogue of 1845 – but fortunately it was accurately drawn by Owen's illustrator George Scharf, and is now recognized as being a second mandible of *Mylodon darwinii*.

Other fossils identified as *M. darwinii* by Owen include half of another mandible, whose teeth he sectioned to observe their internal structure, and parts of a forearm bone (humerus) and thigh bone (femur). Neither of these bones is mentioned in Owen's description of the *Beagle* fossils, but they are listed in his 1845 catalogue of the Royal College of Surgeons collection, probably because by then he had taken delivery of a complete skeleton of *Mylodon* sent from Buenos Aires by the redoubtable Woodbine Parish, and used it to identify Darwin's specimens.

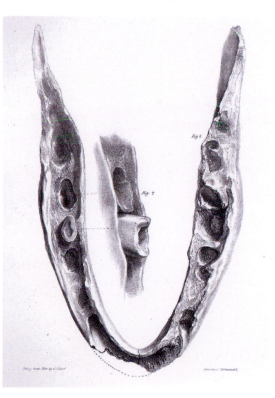

The sloth mandible, now lost, that Owen ascribed to the North American *Megalonyx* but since identified as *Mylodon*. The single remaining tooth can be seen on the left and in the inset.

Based on this skeleton and subsequent finds, we now know that *M. darwinii* was a large ground sloth, estimated to have weighed between 1.5–2 tonnes, similar to a living rhinoceros. It was evidently adapted to cold climates as its range extended to the southernmost part of the continent. There, in the dry conditions of Ultima Esperanza ('Last Hope') Cave in Chile, remarkable finds of its skin and dung were made in the 1890s. The site is now known as Mylodon Cave and the remains are dated to 14,000–12,000 years ago. They show that the animal had a thick pelt, the skin of which was impregnated with pebble-like bony nodules, presumably as

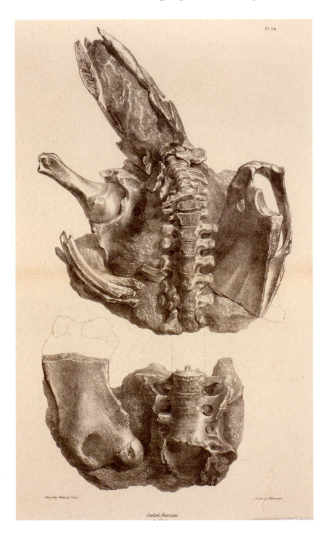

The most complete fossil mammal found by Darwin – the skeleton of the ground sloth *Scelidotherium*. The underside of the skull is at the top. The specimen was approximately 4 ft (1.3 m) long when drawn; some bones had already been removed.

protection from predation. Analysis of pollen preserved in the dung has indicated a diet of grasses, sedges and herbs, probably in a largely treeless environment, at least in this southern part of the animal's range. The discovery of these remains led to speculation that the animal might still be extant, with several purported sightings, but the search has long been called off.

The most complete fossil of any mammal discovered by Darwin was an almost entire skeleton, later recognized by Owen as another new species of ground sloth. It appears to have been found on the beach, partially embedded in loose sand, and Darwin concluded that the whole mass had fallen from the cliff. The skeleton comprised the skull, vertebral column, ribs and limb bones down to the claws, 'all nearly in their proper relative positions', including even the knee-caps. This remarkable discovery was first noted by Darwin on 1 September 1833, and must have been made at some point during his second phase of collecting at Punta Alta the previous week. He quickly recognized the significance of finding an articulated skeleton; whereas odd bones might have been washed out of earlier deposits or fallen in from above, a complete skeleton embedded in sand demonstrated conclusively the contemporaneity of the living animal with the ancient deposit in which it was found. 'Gran bestia all nonsense' he exclaimed in his notebook, for it was perfectly clear that the remains were ancient and not those of a mysterious creature that still roamed the Pampas. He wrote to his sister Caroline that he had discovered the skeleton of an animal 'of which I do not think there exists at present on the globe any relation'. Later he wondered if it might represent the same species as the mandible whose four teeth he had illustrated, subsequently named *Mylodon darwinii*. On detailed comparison, however, Owen confirmed it as a different genus, smaller and with a more elongate skull, and named it *Scelidotherium leptocephalum*.

The ground sloth *Scelidotherium*, with a long, low body and snout. Recent evidence suggests it may have been a burrower.

The lightest of the sloth species discovered by Darwin, *S. leptocephalum* still weighed in at nearly a tonne, equivalent to a large buffalo. It had a relatively long, low body, and its narrow snout suggests quite selective feeding. The remarkable evidence that it was also an efficient burrower is discussed below (see p.51).

The fourth sloth species discovered by Darwin was found in November 1833 during his two-week excursion across present-day Uruguay. It was part of the back of a skull, later named *Glossotherium* by Owen. The find was made in the same stream, the Sarandi, where the larger, more complete skull of the celebrated large mammal *Toxodon* had been discovered (see p.76). It is not quite clear whether Darwin himself found the *Glossotherium* specimen at the stream, or obtained it from the finder together with the *Toxodon*, although the former seems more likely.

At any event the specimen, which Darwin described as from 'an animal rather larger than the horse', was remarkable for its superb state of preservation; he wrote that it appeared 'so fresh that it was difficult to believe [it had] lain buried for ages under

The back part of a skull, 8 in (20 cm) long, of *Glossotherium* collected by Darwin in present-day Uruguay. Top: side view with cheek bone (bearing specimen numbers) and to the right the rounded condyle where the skull attached to the backbone. Below: the inner view of the skull, including the braincase, and showing preservation of delicate bony structures.

ground'. Not only was the appearance of the bone fresher than any of his other fossil finds, it preserved delicate parts that are usually broken away in ancient remains. This included the tympanic bone, one of the tiny ear bones, its preservation in place in the skull leading Owen to praise 'the care and attention devoted to his specimens by their gifted discoverer'. Darwin wanted to know more. He held a piece of the bone in the flame of a spirit-lamp, finding that it not only burnt with a small flame, but 'exhaled a very strong animal odour'. He sent a piece to Trenham Reeks at the Museum of Economic Geology in London, who had undertaken chemical analyses of several of his rock samples, asking what percentage of animal matter it contained. By this he meant organic material aside from bone mineral, and the answer was 7%. We would now recognize that around a quarter of the original protein content had been retained. The remarkable state of preservation of this skull, and its different appearance from others in Darwin's collection, make it very likely to have fallen from a higher, later level in the river bank than the *Toxodon* and glyptodont remains found nearby.

Having only a fragment of skull at his disposal, Owen was characteristically cautious and identified it as an edentate without specifying to which group it belonged. A large attachment surface for the bone supporting the tongue, and a wide hole for the nerve supplying the latter, led him to reconstruct a very large tongue and to devise the name *Glossotherium* (tongue-beast). He later abandoned the name, considering the skull to belong the same species as the jaw that he had named *Mylodon darwinii*. It is now recognized as being distinct, however, so Owen's name has been reinstated and the species is known as *Glossotherium robustum*. Owen considered the animal might have been an insect-eater, breaking open termite nests like an anteater, but it is now known to have been herbivorous in its habits. Its wide muzzle suggests unselective bulk-feeding on grasses and low-growing herbs. In 2017, based on collagen protein extracted from Darwin's *Glossotherium* skull, a radiocarbon date of around 12,660 years ago was obtained. This is one of the latest known records of the genus, close to the time of its extinction.

Glossotherium robustum had an estimated body weight of around 1.5 tonnes. In spite of this, a remarkable recent suggestion is that *Glossotherium* and/or *Scelidotherium* may have constructed large burrows to escape predation or unfavourable weather. Several lines of evidence support this idea. First, several large 'fossil burrows' have been discovered, especially in the area around Buenos Aires, their diameter of 3–4½ ft (1–1.5 m) matching the body width of these species. Second, the forelimb bones of these animals appear modified for very powerful movements like digging. Third, claw marks have been found on the inside of some of the burrows, forming

Glossotherium, the third new species of ground sloth discovered by
Darwin, named by Owen for its supposedly long tongue.

pairs of grooves that match closely the claws of the large second and third digits
seen in these species. These sloths would be by far the largest animals known to
burrow in this way – and one of the fossil burrows is more than 130 ft (40 m) in
length.

Darwin's discovery of four genera of large ground sloths was remarkable,
and also serendipitous in that the area in which he was collecting happened to
be the only region where all four could have been found together. *Mylodon* is
distributed in the southern half of the continent, *Glossotherium* in the northern
half, and *Scelidotherium* in the middle. The genus *Megatherium* is widespread, but
M. americanum is known mainly from Argentina. Only in the Pampas region and
La Plata basin do they overlap. The differing forms of their skulls, and teeth and
limbs, described above, show how several species could have co-existed in the
Late Pleistocene, using different food and habitat resources.

For Darwin, the relationship of the extinct giant sloths to the living species was
one of the examples that led him to his 'law of succession of types', whereby there
was an affinity between the past and present inhabitants of a particular region
(in this case South America). This general pattern was one of the key factors that
ultimately persuaded him of the reality of evolution (see Chapter 6). We now

know that sloths have a fossil record extending back some 35 million years, with around 50 fossil species described. Most were ground-dwelling and of medium to large size, so the six living species are in many ways the unusual ones. The species discovered by Darwin and his immediate predecessors fall into two large groupings: one comprising *Mylodon, Scelidotherium* and *Glossotherium*, the other including *Megatherium* and possibly the North American *Megalonyx*. Living sloths comprise two groups, the two-toed and three-toed sloths, the names reflecting the number of digits visible in the front foot. But the relationships among the living and fossil sloths remain unresolved. It is evident, however, that major changes of body size and adaptation have occurred in the evolution of this remarkable group of animals.

Living tanks

On 22 September 1832, the first day he collected fossil mammal remains in South America, Darwin wrote in his notebook that he had found teeth and a thigh bone (whose identity, from that scant description, remains uncertain), and then one word – 'armadillo'. Almost certainly this was part of the bony carapace of a glyptodont, an extinct South American family and one of the most bizarre groups of mammals ever to evolve. Darwin was to find similar remains on at least four further occasions, and their evident relationship to the living armadillos provided another example of his law of succession of types. This was not, however, before some twists and turns in understanding the identity of the strange dermal armour.

Darwin was well primed to identify for himself the fossil found on that first day, for only a week before, after a day's hunting on horseback at the invitation of some local gauchos, he had feasted on armadillos roasted in their shells. There are 21 living species of armadillo, and they all share a feature unique among living mammals – buried in the skin beneath their outer scaly case is an inner shell made of hundreds of small, interlocking bony plates. The bony shell of a glyptodont is unmistakably similar, just very much larger. When, a few weeks later, Darwin made a second discovery of glyptodont shell at Punta Alta, he aptly described it as an 'armadillo on a grand scale'. His find comprised two sheets of 'thick, osseous polygonal plate, forming a tessellated work', but Darwin recognized that the two sheets were parts of a single hemispherical carapace, squeezed flat by the weight of sediment after burial. Wedged between them he found a small piece of a foot bone. Each sheet measured 3 x 2 ft (90 x 60 cm), of which he managed to recover and send home several fragments of around 6 in (15 cm) in length.

A living armadillo. The claws are used for digging; the keratin scales are underlain by a bony shield similar to that of glyptodonts.

Darwin next encountered glyptodont remains a year later, during his long ride with armed escort from Bahia Blanca to Buenos Aires. On 19 September 1833 they passed the town of Guardia del Monte, and on the shore of a nearby lake Darwin found a 'large piece of tessellated armour'. When, a few weeks later, he visited the Sarandi stream (in present-day Uruguay) where his *Glossotherium* skull was discovered, he picked up further pieces of the characteristic bony shell from the stream bed. At another locality in the same region, fragments of carapace were associated with a shin bone (tibia), later identified by Richard Owen as matching that of a glyptodont.

The most spectacular find, however, was seen on 10 October 1833, at the Tapas stream near Bajada de Santa Fe, where an almost complete carapace of a glyptodont was exposed. Buried in the ground was a bony shell 4½ ft (1.5 m) in diameter, 'the inside of which, when the earth was removed, was like a great cauldron'. It seems likely that this was not a chance encounter but that Darwin was deliberately taken to the spot, as he recorded that the animal had been found a few years previously and all the bones of the skeleton had since been removed, leaving only the case. The bone of the shell was by now quite soft, but Darwin noted its smooth internal surface.

Some of the fossils that influenced Darwin were seen not in the field, but in the collections of people he met. At the house of a priest near Montevideo he was shown a portion of tail some 18 in (46 cm) long. He found it remarkable, since the tail vertebrae were encased in a tube of bony plates, rendering it extremely solid and heavy. He later noted that it 'precisely resembled, but on a gigantic scale, that of the common armadillo'.

Darwin at first had little doubt what he had discovered. In a letter to John Stevens Henslow in October 1832 he wrote of the pieces found at Punta Alta: 'Immediately I saw them I thought they must belong to a giant armadillo, living species of which genus are so abundant here.' This view appeared to be confirmed

Two parts of the glyptodont carapace, *Neosclerocalyptus*, found by Darwin at Punta Alta, each 2 in (4.5 cm) across. The individual bony tesserae can be seen. Between them are two views of the animal's bony claw.

by his reading of the Jesuit traveller Thomas Falkner, whose travel memoir was in the *Beagle* library. Sixty years previously, Falkner had become the first person to describe the glyptodont's bony shell, based on a 10 ft (3 m) dome he had unearthed in Patagonia; he too likened it to a huge armadillo. However, Darwin's instincts were soon to be confounded when he learnt that the prevailing view, promoted by the revered Georges Cuvier and others, was that the bony sheets were in fact the armour of a very different creature, the giant sloth *Megatherium*.

This misapprehension rested on flimsy evidence. Cuvier had first proposed it in his *Ossemens Fossiles* of 1823, quoting a letter from a priest based in Montevideo who had described the finding of a partial skeleton with the characteristic bony armour. The correspondent had named it *Dasypus* (i.e. armadillo), but then added in brackets, 'Megatherium'. When William Clift, Owen's superior at the Royal College of Surgeons, received three skeletons sent by Woodbine Parish from Buenos Aires in 1832, two of which included portions of bony armour, he followed Cuvier and assigned them to *Megatherium*. When Darwin first became aware of this theory, he wrote to Henslow to ask what the evidence for it was – he was evidently puzzled. From that point on, entries in his notebooks, diaries and letters show him wavering between trusting his own instincts, and submitting to the word of authority. At one extreme, referring to the cauldron-like shell he had seen at the Bajada de Santa Fe, he described it as a 'paluda's case', paluda being the Spanish name for one of the larger species of living armadillo, a specimen of which Darwin had described and sketched in his notebook. In his *Geological Diary* of October 1833, however, he pondered the issue and persuaded himself Cuvier's idea was tenable since *Megatherium* bones had been found in beds of the same age as those containing the bony shell. Eventually, in November 1833, he came to a compromise that both described the shell accurately and noted its supposed affinity: it was the 'armadillo-like case of the Megatherium'.

It was Richard Owen who, in 1841, finally laid the question to rest. From detailed observation of Parish's specimens, Owen showed conclusively that the bones associated with the carapaces closely resembled those of the armadillo and not *Megatherium*. Moreover, by now he could list no fewer than 12 partial skeletons of *Megatherium* known to science, none of them associated with a bony carapace. Owen was not the first to notice that bones associated with the bony armour were similar to those of armadillos, but he was the first to conclusively dismiss the *Megatherium* theory and to show that the armadillo-like giant was a genus in its own right, which he named *Glyptodon*.

Skeleton of the largest glyptodont, the 10 ft (3 m) long *Glyptodon clavipes*. The animal was protected by its rigid bony carapace, head shield, and its flexible bony tail.

Perhaps because they were so bizarre, the features shared by the extinct glyptodonts and living armadillos provided an example of the law of succession of types that struck Darwin forcefully (see Chapter 6). It is ironic, therefore, that he was slow to recognize that the issue of the carapace had been resolved in his favour. In his 1844 essay on evolution, the precursor to *The Origin of Species*, Darwin still referred to some species of giant sloths possessing bony armour like that of the armadillo. As historian Sandra Herbert has discovered, an undated slip among Darwin's notes reads 'Owen's paper on Glyptodon must be studied'. At last, in 1846, he described his find from Punta Alta as 'the bony armour of a large dasypoid quadruped' (dasypoid meaning that it actually was an armadillo, not just with a similar shell), and the one from Guardia del Monte as 'a large piece of tessellated armour, like that of the Glyptodon'. Once the link between armadillos and glyptodonts had been clarified in his mind, the example became a key part of the evidence for the succession of types, and ultimately for the theory of evolution.

Modern research has amply confirmed the close relationship between glyptodonts and armadillos, but has also highlighted their differences. Size is the most obvious – the largest living armadillos are around 30 in (75 cm) long excluding the tail, and weigh around 66 lb (30 kg), but most glyptodonts weighed several hundred kilograms, the largest reaching 2 tonnes in weight and 10 ft (3 m) in length, the size of a small car. Since their closest armadillo relatives appear to include some of the smallest living species, it has been estimated that their common ancestor weighed only 13 lb (6 kg) – emphasizing the spectacular size increase in the evolution of the glyptodonts. Further differences relate to the shell: while that of the armadillo is jointed, allowing flexibility and the ability to roll up into a ball when threatened, the glyptodont carapace was a completely rigid box. When Darwin saw part of a glyptodont tail, he described it as 'a most formidable weapon', a supposition strongly confirmed by subsequent research. Not only was the tail spiked or clubbed in some species, but calculations suggest that its blow was powerful enough to crack the carapace of another glyptodont. In their diet, glyptodonts are believed to have been strictly herbivorous, and while living armadillos will take some plant matter, they mainly eat insects and other invertebrates.

The question remains – which species of glyptodont did Darwin discover? Recent studies using DNA recovered from well-preserved fossil bones show that the glyptodont family emerged from within the armadillo group some 35 million years ago. From that point on, dozens of different glyptodont species evolved, some eight of which occur in the Late Pleistocene deposits where Darwin unearthed his treasures. Darwin himself realized there were at least two sorts of carapace among his finds – the pieces he found in 1833 on the Sarandi were thicker and had a different pattern than the sheets dug out a year earlier at Punta Alta. None of Darwin's glyptodont specimens has survived, but from Owen's illustrations the Punta Alta find (see p.55) can be identified as *Neosclerocalyptus*. This medium-sized glyptodont was around 8 ft (2.5 m) long, with an elongated body shape like a flattened cylinder, and has been estimated as weighing 660–1,300 lb (300–600 kg). It has a feature unique among the group – a bulbous expansion of its nasal region, housing a network of bony sinuses; in life these would have borne membranes reducing the loss of heat and moisture as the animal breathed. These features, similar to those found in mammals such as the saiga antelope of central Asia, are adaptations to cold, dry environments, and correspond to the distribution of *Neosclerocalyptus* only in the southern part of South America, and a particular abundance of its fossils in the coldest part of the last glaciation. Accordingly, the skulls show the animal had a wide muzzle – an adaptation to bulk-feeding in open grassland, where it ate both grasses and herbs.

The identity of the other Darwin specimens is less certain, since they were never illustrated. Owen's description of the Sarandi fragments, including their great thickness of nearly 1½ in (4 cm), makes it likely that they were a species of *Glyptodon* (the genus that gives the group its name), either *G. reticulatus* or *G. clavipes*, with a body mass of up to 2 tonnes. *Glyptodon* had a narrower snout than *Neosclerocalyptus*, indicating that it fed more selectively on high-nutrient

Neosclerocalyptus, the glyptodont identified from Darwin's fossil fragments. With a flattened, elongated shell and a wide muzzle, it was adapted for feeding on open grassland.

plants. The *Glyptodon* carapace was roughly hemispherical, and the spiked tail armour was jointed instead of being a rigid tube as in other glyptodonts. It has recently been discovered that *Glyptodon* had a further line of defence – rows of bony spikes around the rim of its carapace. On the identity of the more complete 'cauldron' seen by Darwin at the Tapas stream we cannot even speculate, as it could correspond to almost any glyptodont species.

Wild horses

Although Darwin was thrilled by his discovery of huge bones and skulls of extinct mammals, nothing caused him more excitement than the finding of a humble horse's tooth. It was always the scientific significance of a discovery, rather than its magnificence or even novelty, that drew his keenest interest. At Bajada de Santa Fe in October 1833, in the red clayey deposit of the Tapas stream where the glyptodont carapace lay, Darwin found many other bones exposed. These included a single molar tooth of a horse – a find that, unlike many of the others, he could immediately identify with certainty. It appeared to be embedded in the same layer as the remains of extinct fauna, but since fossil horses were at that time unknown in South America, Darwin doubted the evidence of his eyes. Might it have been washed down from close to, or even on, the modern land surface, muddy sediment then hardening around it? Was it, in other words, the worthless tooth of a domestic horse post-dating the Spanish

The fossil molar of a horse, 3 in (7 cm) long, in the Darwin collection, believed to be the specimen from Bajada de Santa Fe from which Darwin first documented native horses in South America.

conquest? After careful scrutiny, Darwin satisfied himself that the tooth had indeed originated within the same ancient layer as the extinct beasts. Back in London, he and Owen agreed that the state of preservation of the horse tooth was identical to that of the extinct mammal *Toxodon*, found nearby – both were similarly corroded and stained red. To strengthen his argument still further, Darwin noted that the surrounding country was uninhabited and without fresh water, so modern domestic animals were unlikely. He need not have worried; it later emerged that a similar horse tooth, albeit somewhat less well preserved, lay embedded in a sediment block he had collected alongside bones of giant sloths and *Toxodon* at Punta Alta two months earlier.

Darwin marvelled at the finds, discussing them at length in his notebooks, in the 1839 *Journal of Researches,* and even in *The Origin of Species* published a quarter of a century after the discovery. With characteristic English understatement, Richard Owen described the finding of the horse teeth as 'not one of the least interesting fruits of Mr. Darwin's palaeontological discoveries'; but celebrated US palaeontologist George Simpson later called it 'the most important single result of Darwin's collections of fossil mammals during the voyage of the *Beagle*'. The cause for Darwin's astonishment lay not only in the co-existence of extinct species with those still alive, for this had already been demonstrated in the bone caves of Europe, but in the revelation that horses had once lived wild in South America, and the puzzle it posed about the causes of extinction. It was well known that the Spanish had found no horses when they arrived in the Americas at the turn of the 16th century yet, as Darwin emphasized, once introduced and escaping into the wild, horses had multiplied in huge numbers and were clearly well adapted to the Pampas environment. So if they had previously existed there as native animals, why had they gone extinct? An apparent solution did not present itself until, back in London, Richard Owen studied the remains.

In his first description of Darwin's fossils, published in 1838, Owen considered the horse teeth to be similar to those of living horses, but reserved judgement, calling them only 'a species of horse'. By 1845, however, he had decided it was a previously unknown species, which he named *Equus curvidens*, based on the distinctly curved shape of the tooth crown seen from the side. This changed Darwin's perspective on the find, to the extent that, recalling his earlier reaction, he exclaimed in *The Origin of Species*, 'How utterly groundless was my astonishment!' He now reasoned that if it were a different species from the living horse, then its habitat and diet could have differed too, even if they were essentially unknown to us. It was therefore easy to imagine that if conditions had changed so as to become less favourable to the species, it might have been driven to extinction. The new conditions could, however, have been favourable to the success of the modern species once it was introduced in historical times.

Since Darwin's day, fossil horses have been found abundantly in South America. Darwin's specimens are now named *Equus neogeus*, a name given by the Danish naturalist Peter Wilhelm Lund to a leg bone of a horse he had unearthed in the late 1830s in a cave in Brazil, and now considered the same species. By the rules of zoological nomenclature, Lund's name of 1840 takes precedence over Owen's of 1845. The species has also been given the name *Amerhippus*, often in the combination *Equus* (*Amerhippus*) *neogeus*, the meaning ('American horse of the New World') leaving no doubt as to its geographical origin.

Study of its remains has indicated that *E. neogeus* weighed some 660–830 lb (300–375 kg) and stood around 4½ ft (1.5 m) at the shoulder – about the size of a light riding horse but quite large for a wild species. Chemical analysis of the fossil bones indicates that the horses ate primarily grasses, but included a variety of grass species in their diet, depending on locality.

The extinct South American wild horse, *Equus* (*Amerhippus*) *neogeus*, named *Equus curvidens* by Owen on the basis of Darwin's fossil tooth.

Horses had entered South America when the isthmus of Panama formed, connecting North and South America, around three million years ago. *Equus neogeus* was one of several species that diversified there, each occupying a somewhat different area of the continent, although their ranges overlapped. *Equus neogeus* lived along the eastern seaboard of the continent, from the Argentinian Pampas (where Darwin found its remains) in the south, to the northeast corner of Brazil some 3,000 miles (5,000 km) to the north. Recent genetic evidence, based on DNA extracted from fossil bones, has shown that the species was very closely related to the living Przewalski's horse of Mongolia, the last representative of Eurasian wild horses, and hence to the domestic horses that derive from them. Remarkably, therefore, horses of the late ice age may have formed an almost continuous population stretching from Europe across Asia and into North and South America. In turn, this discovery calls into question whether the original wild horses of South America were so different from the domestic horses that have replaced them, and reignites Darwin's original question – why did they go extinct? We will return to this question in Chapter 6.

It is one of many coincidences characterizing Darwin's *Beagle* discoveries that on 7 September 1833, only a month before his excavation of the horse tooth, he had seen a soldier striking a fire with a piece of flint, and was told it had been found with many others at a crossing of the Rio Negro. Darwin recognized it as a prehistoric arrowhead, barbed and unlike anything used by local people at the time of his visit. He immediately speculated that a change in the method of catching animals must have taken place with the introduction of horses – from hunting on foot with bows and arrows, to the lassos and balls he had seen being thrown with great skill to catch animals from horseback. Pondering these facts some time after the discovery of the fossil tooth from Bajada, he again doubted whether horses had really been native to South America, until the discovery of a second tooth, in sediments from Punta Alta, 500 miles (800 km) to the south, laid the matter to rest.

Giants' bones

On their ride from Buenos Aires to Santa Fe, Darwin and his companions lodged near the town of Rosario for the night of 30 September 1833. During the night Darwin's pistol was stolen so, feeling unsafe, the party left by moonlight, arriving at the Carcarañá River at sunrise on 1 October. Darwin had read the account of an earlier English traveller, Thomas Falkner, who in the 1770s described having seen huge bones in the banks of this river, a tributary of the Paraná, and Darwin spent the whole day fossil-hunting in the cliffs. At one point, locals informed him that 'giants' bones' were to be seen a short distance away on the western bank of the Paraná itself. The remains could only be reached from the river, some 1,000 ft (300 m) wide at this point, so Darwin hired a canoe and was guided to the spot. There he found, about 6 ft (2 m) above the water, two gigantic skeletons projecting from the face of the cliff. This must have been an astonishing sight, for the bones, in Darwin's words, 'still held their proper relative positions'. In other words, they were complete or partially complete skeletons of two huge animals, not just bits and pieces of various individuals. Local people told Darwin they had long known of the skeletons, and had wondered how they had got there, concluding that they were the remains of a giant burrowing animal! Similar theories had arisen in Siberia when natives uncovered the remains of frozen mammoths.

The skeletons seen by Darwin were, however, badly decayed – he described them as 'soft as cheese', and was able to collect only a few small fragments. He realized that they lay below the high-water mark of the river, and so had been subject to repeated submergence and drying out, accounting for their poorly preserved state. Even the fragments he collected are now lost, but Darwin's identification of them as mastodon was almost certainly correct. The mastodon, a distant relative of the elephant, was one of only two of the giant mammals found by Darwin whose existence was already known at the time (the other was *Megatherium*). His description of the bones as 'immense' and 'gigantic' is the first clue, since the limb bones of mastodons are much larger and more robust than any of the other South American fossil mammals of this age. Further, Darwin explicitly mentions that their massive molar teeth were in place, and even though they fell to pieces in his hands, their form was clear. The molars of a mastodon, some 8 in (20 cm) long, with large rounded cusps and enamel up to a centimetre (half an inch) thick, are unmistakable. Darwin may well have seen mastodon teeth while a student at Cambridge, and he mentions several occasions on the *Beagle* voyage when he was shown mastodon teeth by collectors. He also had for reference, in the *Beagle* library, a 15-page description of mastodon fossils, with illustrations by no less an authority than Georges Cuvier.

The two great skeletons were, moreover, not the only remains of mastodon Darwin found that day. On the banks of the Carcarañá he found an isolated tooth, also decomposed but similar in form to the one from the Paraná. A third

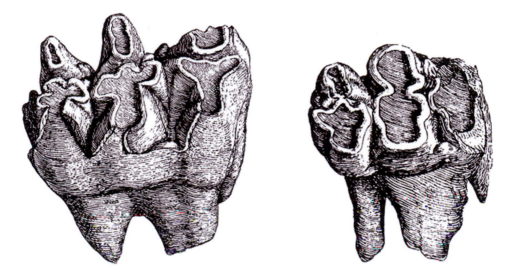

Teeth of the mastodon *Cuvieronius hyodon*, collected by
Alexander von Humboldt and illustrated by Georges Cuvier in 1806.
Drawings such as these, in books held in the *Beagle* library, helped
Darwin to identify his finds.

decaying molar – more of an impression than an actual fossil – was found nine days later at Bajada de Santa Fe, on the other side of the Paraná, alongside the glyptodon carapace and teeth of horse and *Toxodon*.

The name mastodon, meaning 'breast-tooth', had been coined by Cuvier in 1806, the smaller cusps on the molars apparently reminding him of human nipples. He had received fossils of the best-known species, the North American mastodon, from the famous site of Big Bone Lick in Kentucky, but also had access to remains from Europe and from South America, which he recognized were of the same general kind, although differing in detail and therefore representing distinct species. His key specimens from South America had, in fact, been collected by Darwin's great hero, Alexander von Humboldt, during his expeditions in 1799–1804. In Colombia in particular, at a place known as the 'Field of the Giants', Humboldt had reported seeing an immense accumulation of bones. It is revealing that Darwin, writing in his *Geological Diary* soon after the discoveries on the Paraná, wrote 'I believe it to be the narrow-toothed mastodon', a direct reference to the name applied by Cuvier to South American fossils he had studied. Darwin can have gleaned this only through reading Cuvier's works – an insight into his scholarly methods while on the *Beagle*.

When Richard Owen examined Darwin's mastodon fossils in London, he acknowledged that from such fragmentary material, identification of the precise species was difficult, but he was able to state (correctly) that they did not pertain to the North American mastodon, which has sharper cusps. We now know that the latter species, *Mammut americanum*, belongs in an entirely different family from the South American species, which are part of the large grouping known as gomphotheres. The term 'mastodon' to encompass all of these diverse elephant relatives is now used only informally.

When Owen came to write a catalogue of the Royal College of Surgeons collection in 1845, he listed one of Darwin's tooth fragments as '*Mastodon andium*', a name that had been given by Cuvier to a tooth found by Humboldt in the Andes of Ecuador. Darwin himself commented that if the same species extended from the Pampean plains to the high mountains, its ecology was difficult to define. This puzzle has since been at least partly resolved, since the Pampean fossils are now recognized as being a different species from those of the Andes. The former are now called *Notiomastodon platensis* after the Rio Plata, the latter *Cuvieronius hyodon* after the great anatomist. Since *Notiomastodon platensis* is the only proboscidean species now recognized in the region of modern-day Argentina where Darwin found his fossils (see map opposite), it is very likely that this was their true identity.

Since Darwin's day, gomphothere fossils have been unearthed from many more localities across South America. Their apparent variation has led to differing opinions on the number of species that existed – at one time as many as seven species were recognized. Currently most researchers consider that only *Cuvieronius hyodon* and *Notiomastodon platensis* are valid, although it is acknowledged that the latter lived over a very wide area (see map) and was rather variable. It is possible that further research may show fossils from Brazil and the northwest of the continent to be a different species from *N. platensis* of the Pampas.

However many species are represented, it is clear that gomphotheres arrived relatively recently in South America. With a history in North America spanning at least 12 million years, they entered South America only after the closure of the Panama isthmus around three million years ago, along with other immigrants such as horses and big cats. *Cuvieronius* has been found at sites in southern USA and Mexico, the earliest dating to around five million years ago, so very likely included the direct ancestors of South American *C. hyodon*. But *Notiomastodon* has not been found in North America, and the earliest known fossils are about

RIGHT Distribution map of *Notiomastodon platensis* (orange diagonals) and *Cuvieronius hyodon* (orange), based on all known finds. The location of Darwin's mastodon fossils is shown by a star.

BELOW The gomphothere *Notiomastodon platensis*. This distant relative of today's elephants stood around 8 ft (2.5 m) at the shoulder and weighed around 4 tonnes.

500,000 years old, so it has generally been considered to have evolved from *Cuvieronius* in South America. Recent research, however, suggests that it arose from a different North American ancestor, in which case the two gomphotheres would have migrated separately into South America.

Gomphotheres resembled elephants in general build but were shorter and stockier. Of the two South American species, *C. hyodon* was somewhat smaller in build and its upper tusks were twisted into a lyre-shape, while those of *N. platensis* were gently curved. The latter species stood around 8 ft (2.5 m) at the shoulder and weighed around 4 tonnes, similar to a living Asian elephant.

Most fossils of *N. platensis* are from lowland areas, below 3,300 ft (1,000 m) although a few show that they occasionally ventured up to higher altitude. In general, the species seems to have favoured dry, lightly wooded habitats. In a remarkable study published in 2012, Brazilian scientists were able to determine the diet of *N. platensis* by applying new techniques to teeth found in a cave in Brazil. First, they examined the chewing surface of the tooth enamel microscopically and found extensive scratching. When compared with various living mammals this indicated quantities of grass in the diet. Next, they scraped calculus from the side of the tooth and dissolved it with acid. The residue contained tiny plant fragments that had become trapped in the growing calculus while the animal was feeding. Under the microscope the scientists identified pollen grains and wood fragments of shrubby species, showing that the animals had also eaten leaves and twigs. Such findings would surely have delighted Darwin who, with the limited knowledge available at the time, was always cautious in speculating about the ecology of extinct species.

Gnawers and burrowers

The cliff at Farola Monte Hermoso, some 30 miles (50 km) east of Punta Alta in the bay of Bahia Blanca, has a geological stage named after it, the Montehermosan. Its significance rests chiefly in its rich fossil fauna, and the first naturalist known to have collected fossils there was Charles Darwin during the *Beagle* voyage. Circumstances, however, severely limited the amount of material he was able to procure.

Only a week after Darwin's first successful fossil-hunting excursion at Punta Alta, the *Beagle* anchored off Monte Hermoso and a party was sent ashore to erect a marker for the ship's surveying work. Darwin went with them to search for fossils, and FitzRoy recorded in his journal that another of the party 'discovered many curious fossils in some low cliffs'. Nothing is known or was preserved of their finds that day, however, for as recounted (see p.21), the party were soon more concerned with surviving the night in the midst of a raging storm, with no food and little shelter.

Two weeks later, on 19 October 1832, they were back at the same spot (now named 'Starvation Point' by the crew), and Darwin accompanied Fitzroy onto shore, but only for half an hour. In this limited time he sketched the geology of the cliff, collected a few fossils, and noted the occurrence of many more.

The fossils from Monte Hermoso were strikingly different in appearance from other mammalian bones Darwin had collected. They were, he noted, heavy, very hard, dark red to black in colour and with a surface polish. A palaeontologist today would immediately suspect that they were of greater geological age, although Darwin attributed these features to prolonged immersion in water before burial.

Farola Monte Hermoso, Argentina, named Starvation Point by the crew of the *Beagle*. In 1832, Darwin collected fossils of rodents and other small mammals from the cliff close to beach level.

He vacillated in his opinion of their age, considering them at first of similar age to the finds from Punta Alta, but later deciding they were somewhat older. He based this, firstly, on the higher elevation of Monte Hermoso – the land had been uplifted for longer. Secondly, he found a single fragment of bone at Punta Alta that was 'black as jet' and resembled those from Monte Hermoso, but it was 'very much rolled' – an indication that it had been washed out of an older deposit at the time of formation of the Punta Alta beds and incorporated into them – a process now known as reworking. He nonetheless considered the difference in age to be relatively small and anticipated that fossil species from the two sites would be similar to each other. In large part this was because the fossil-bearing sediments appeared so similar – at both locations they comprised the reddish clays and silts of his Pampean Formation, with areas of hardened rock known locally as tosca.

We now know that the rather uniform 'Pampean' sediments represent some 430 ft (130 m) of cumulative deposition spanning as much as 12 million years of geological time. The Montehermosan, exposed in the lower part of the cliff where Darwin collected, dates to between six and four million years ago. Spanning the end of the Miocene and beginning of the Pliocene epochs (see pp.32–33), this is the age of the mammal remains from Monte Hermoso. In contrast, the finds from Punta Alta and Darwin's other localities are at most half a million years old, and in many cases much less.

The mammalian fossils from Monte Hermoso differed in another way from those he had collected elsewhere – they were the remains of much smaller kinds of animals. Although Darwin reported seeing many bones of larger mammals at Monte Hermoso, including some he identified as those of a giant sloth, all the specimens he collected were those of small mammals. Darwin described how some areas of the deposit were packed with small bones, including almost perfect skeletons. The fossils he collected that day included parts of the jaws and teeth of one animal, much of the hind-foot of another, an isolated molar tooth, and parts of a hip-joint of a fourth individual.

The foot bones gave him an impression of the size of the animal; it was comparable to, but slightly smaller than, the living Patagonian mara, a South American rodent, then named *Cavia patagonica*. Darwin considered the fossil specimens to be closely related to, but distinct from, the living mara. As such, it was his best example of a modern species replacing an extinct one, since in other cases (glyptodont to armadillo, giant to living sloths) there was a relationship but it was less direct. Palaeobiologist Niles Eldredge has suggested that Darwin's

The enigmatic foot bones, 3 in (7 cm) long, collected by Darwin at Monte Hermoso. Long considered those of a rodent, they are now identified as belonging to a different kind of mammal altogether, a distant relative of *Toxodon*.

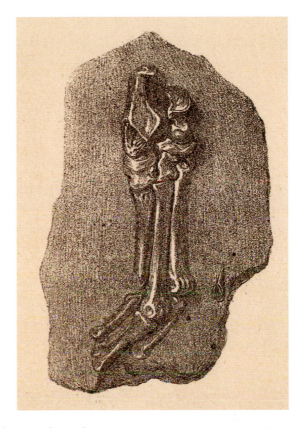

'*Cavia*' fossils were a critical early stimulus in his conversion to transmutationism, going so far as to describe them as 'the most important single fossil species in the history of evolutionary biology', and Monte Hermoso as being 'as important in the development of Darwin's evolutionary thinking as the Galápagos Islands'. This will be further discussed in Chapter 6, but it is clear that the little bones were of significance to Darwin as, nearly a year-and-a-half after their discovery, he referred to them in a letter to Henslow, particularly concerned lest the labels indicating their provenance be lost.

Owen, in the event, was not able to determine which kind of animal the fossil foot bones represented, beyond considering them to be some kind of rodent. He was, however, more confident with the jaw bones, which from the number and form of the molars he identified as a species of the genus *Ctenomys* – smaller, burrowing rodents commonly known as tuco-tucos, from the 'tuc-tuc' sound they make while digging their tunnels. He noted some minor differences from the living species, however, and later named it as an extinct species, *Ctenomys antiquus*. In his 1839 *Journal of Researches*, Darwin simply switched genera in his implication of a living species replacing an extinct one – instead of *Cavia*, the jaw bone was 'part of the head of a Ctenomys; the species being different from the Tucutuco, but with a close general resemblance'. He was familiar with the tuco-tuco, having kept several as pets during the voyage.

The remaining bones in Darwin's Monte Hermoso collection – hip bones and an isolated molar – were of a considerably larger animal. The creature was as big as the present-day capybara, the world's largest rodent, standing some 2 ft (60 cm) at the shoulder and commonly weighing up to 110 lb (50 kg). Owen concluded that the bones were 'sufficient to prove that there once existed in South America a species [of rodent] as large as the present capybara, but now apparently extinct', a conclusion that Darwin again quoted with approval.

LEFT The tuco-tuco, a burrowing South American rodent that spends the majority of its life underground. Some 60 species are known, all in the genus *Ctenomys*.

BELOW LEFT AND RIGHT The mara or cavy, *Dolichotis patagonum*, a rabbit-like South American rodent related to the guinea-pig, and the capybara, *Hydrochoerus hydrochaeris*, the world's largest rodent, with a flycatcher bird on its back.

The rich fossil mammal fauna from Monte Hermoso has since been studied in detail. Among 10 different rodent species identified, the only one that belongs in the same family as the living capybara is a species now called *Phugatherium cataclisticum,* and this is very likely the identity of the large molar and hip-bones collected by Darwin. Despite being some five million years old, the appearance and behaviour of this species are believed to have not been very different from those of modern capybaras (*Hydrochoerus*). The body-shape is similar, and rapidly formed fossil deposits with individuals of all ages may indicate that they lived in herds like today's animals. Finally, the fossils have always been found in deposits laid down by water, suggesting that, as with the living species, the ancient capybaras were semi-aquatic. Darwin had surmised as much, when he envisaged that 'a stream (in which perhaps the

LEFT AND BELOW The fossil jaw bones and teeth of a rodent collected by Darwin at Monte Hermoso. Recognized by Owen as related to the living tuco-tuco, it is now named *Actenomys priscus*.

extinct aquatic *Hydrochoerus* lived)' had carried the carcasses to the place where they were preserved. Like the living species, it probably ate a variety of plants growing in and around the water's edge.

As for the fossil jaw bones, it is likely that the two specimens illustrated by Owen were not, as he thought, upper and lower jaws, but two parts of the same lower jaw bone. Nonetheless, his identification of them as belonging to a relative of the living tuco-tuco is very likely correct – the fossil species is now called *Actenomys priscus*. Recent anatomical studies of *Actenomys* suggest that while its hands and feet were adapted for digging as in the living tuco-tucos (*Ctenomys*), the adaptations were not as extreme, and it lacked features of the jaws and teeth that the living species also use in digging. This suggests that while *A. priscus* dug and lived in burrows, it did not live so fully a subterranean life as the tuco-tucos, which spend up to 90% of their time in a network of tunnels. There are as many as 60 species of *Ctenomys* across South America, ranging in weight from 4 oz to 2¼ lb (100 g to 1 kg), but *A. priscus* was larger than any of them at around 2¾ lb (1.3 kg).

The foot bones, by contrast, can now be identified as belonging not to any kind of rodent, but to a different group of mammals altogether. The bones no longer survive, but from Owen's illustration (see p.71) they can be identified as the extinct mammal *Paedotherium*, probably the species *P. typicum*. This animal, common in the Monte Hermoso Formation, belongs to an extinct group of exclusively South American mammals known as notoungulates, among which is the rhino-sized *Toxodon* described on subsequent pages. *Paedotherium typicum*, however, weighed in at only 4½ lb (2 kg) – about the size of a rabbit. Its way of life may have been broadly similar, too, living in burrows and consuming a herbivorous diet. It is unsurprising that Darwin and Owen took the foot bones to be those of a large South American rodent, since with similar habits their anatomy is comparable, and notoungulates (with the exception of Darwin's *Toxodon* skull) were unknown in the 1830s.

The Monte Hermoso mammals are one of the examples where our modern understanding supports Darwin's concept of the succession of types. All of the living and fossil rodents discussed above belong to a large group called the caviomorphs, which are exclusive to South America. The evolutionary history of the caviomorphs goes back some 40 million years, when their ancestors probably arrived from Africa (across the narrow early Atlantic) by rafting. Their closest relatives outside South America are the porcupines. The new

Reconstruction of the extinct South American mammal
Paedotherium. Bones found by Darwin at Monte Hermoso
are now recognized as belonging to this rabbit-sized relative
of *Toxodon*.

identification of the foot bones as those of a notoungulate also fits the idea of an endemic evolutionary radiation, since this group, though now extinct, was exclusively South American (see below).

Whether or not the finding of the small fossil mammals in 1832 represented a 'eureka moment' for Darwin, they played their part in his conversion to transmutationism. The discovery of an extinct but close relative of a modern species, living in the same area, was enough for Darwin to suspect direct descent rather than replacement by a coincidentally similar, separately created species.

'One of the strangest animals ever discovered'

In mid-November 1833, Darwin spent several days at the estate of Mr and Mrs Keen on the Rio Negro in western Uruguay. Mr Keen accompanied Darwin on the first day of his ride back to Montevideo, as he had heard about some remains, believed to be giant's bones, held at a neighbouring farmhouse. Arriving at the estancia they were shown the almost complete fossil skull of a rhino-sized animal, along with some bones of its skeleton. The owner explained that the remains had been found in a nearby stream, the Sarandi, after a flood had washed them out of the river bank. The skull, they said, had been quite perfect when discovered, but by the time Darwin got to it local boys had set it on top of a post for target practice and knocked out the teeth with stones. The lower jaw had apparently also been recovered but had since been lost. Darwin rescued the skull by purchasing it for around a shilling and sixpence, equivalent to some £7.50 today. Thanks to its unusual form and completeness, he regarded it as one of the most valuable finds of his voyage, and today it is among the treasured possessions of the Natural History Museum in London. At the time, Darwin could only guess at its identity, generally describing it just as a large head, or occasionally venturing that it might be the giant sloth *Megatherium* (the only South American species of comparable size known at the time).

The skull, 19½ in (50 cm) long, purchased by Darwin for 18 pence from a Uruguayan farmer. Later named *Toxodon platensis* by Richard Owen, it was one of Darwin's most treasured finds of the *Beagle* voyage.

Darwin also had no idea that some of the fossils he had already found, and that had aroused equal wonder in him, actually belonged to the same species. These were the huge, rodent-like teeth that he had excavated at several localities. He first encountered them during his initial season at Punta Alta, where in September or October 1832 he unearthed the 'molar teeth of some large animal' which, he speculated in a letter to Henslow, seemed to belong to some enormous rodent. If this conjecture should prove right, he wrote in his diary, South America would possess not only the largest living rodent (the capybara), but by far the largest fossil one too. Thanks to the rare preservation of one of Darwin's original

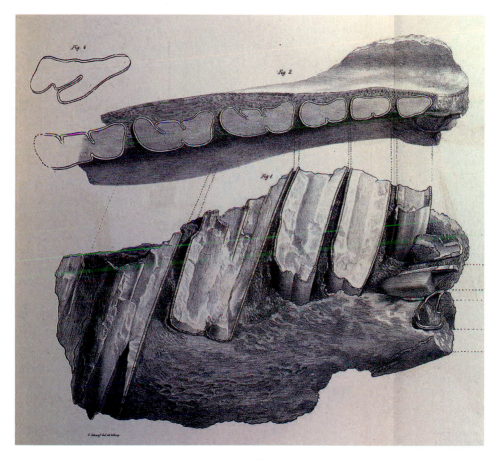

Owen's illustration of the lower jaw of *Toxodon* found by Darwin at Punta Alta. Six molar teeth are visible, as well as three broken incisors at the front (to the right of the picture).

field numbers on a surviving specimen, we know that the Punta Alta find was actually a lower jaw bone, broken into several pieces but containing the remains of 17 teeth – six molars on each side and five incisors. A year later, at the Carcarañá River, another treasure came to light – an 'enormous gnawing tooth' which, as Darwin wrote to Henslow, 'puzzles even my conjectures'. Three further isolated teeth were found: a long, curved incisor at Punta Alta (which he likened to the tusk of a wild boar), a fragmentary upper molar on the Tapas stream and, in the Darwin collection but not mentioned in any of his notes, a large and robust lower molar from an unknown location.

When the great skull was unwrapped in London, Richard Owen was equally impressed. He chose to study it first among Darwin's finds, and gave it pride of place at the start of his monograph on the *Beagle* fossils, devoting 19 pages of closely spaced text to its description. In a brilliant piece of anatomical deduction, he showed that the skull, the lower jaw and the curious rodent-like teeth all belonged to the same creature. He first noticed that the 'enormous gnawing tooth' from the Carcarañá was a molar that fitted perfectly into one of the empty sockets in the skull. Both had the same, strangely scalloped cross-section and there could be no doubt they pertained to the same species, although obviously not the same individual as they were found some 190 miles (300 km) apart.

The 5 in (12 cm) long incisor tooth of *Toxodon*, collected by Darwin at Punta Alta – he likened it to the tusk of a wild boar, while for Owen it suggested an affinity to the rodents.

The *Toxodon* skull illustrated in Owen's *Zoology of HMS Beagle* – though lacking teeth, the shape and position of the empty sockets enabled Owen to link it to the isolated teeth found by Darwin.

The fragmentary tooth from the Tapas stream tied in with this, since it was of similar form and evidently also an upper molar. The lower jaw from Punta Alta was more tricky, but Owen noted that the positioning and spacing of its incisors and molars mimicked those suggested by the empty sockets in the skull, and that the distribution of enamel on the tooth crowns also showed parallels between the isolated upper teeth and those in the lower jaw. Although he exercised some caution in his wording, he was evidently confident of his conclusion, which was soon shown to be correct by more complete finds.

Richard Owen's conclusions about the animal he named *Toxodon platensis* ('curved tooth from the River Plata') are celebrated in the world of palaeontology. As he readily admitted, it bore no close resemblance to anything known, yet comparisons had to be made, and Owen's famous conclusion was that *Toxodon* was 'referable to the Order Pachydermata, but with affinities to the Rodentia, Edentata, and Herbivorous Cetacea'. In modern terminology, he classified it in a group containing other large mammals including elephants, rhinos and hippos (not currently considered to be related), but noting that it also had 'affinities' to animals as diverse as rodents, sloths and sea-cows. To a modern biologist this seems incomprehensible, but we must remember that Owen was working in a pre-evolutionary framework, and that his concept of 'affinity' was different from the modern idea that animals are similar because they are related through common descent. It referred simply to shared aspects of form. Moreover, there was no reason why the Creator might not have taken design ideas from a range of animals when constructing a creature like *Toxodon*, to produce what Owen called a 'chain of affinities' between the different groups; nor did this make the combination of traits in *Toxodon* any less remarkable to Owen himself.

Darwin was enthralled; in the second edition of his *Journal of Researches* in 1845, he exclaimed: 'How wonderfully are the different Orders, at the present time so well separated, blended together in different points of the structure of the Toxodon!' By this time he was already a committed transmutationist, and although he took care in his writings to conceal the fact, in retrospect we can see hints at his thoughts on the matter, which even at the time enlightened readers might have noticed. For him, anything that broke down the perception of rigid boundaries between species or groups of species was grist to his evolutionary mill. And what are we to make of the suggestion that 'at the present time' the Orders were well separated, if not a hint that at some time past there might have been a common ancestor in which the features of now disparate species were merged or combined?

Owen had first described *Toxodon* at a meeting of the Geological Society in London on the 19 April 1837, with Darwin in attendance. At that point, he was inclined to place it within the rodents; Darwin quoted from the proceedings of the meeting: 'Mr Owen says... as far as dental characters have weight, the Toxodon must be referred to the rodent order.' Nor was it lost on either of them that this giant was found in the right place – among the rodents, 'the capybara

Toxodon platensis, the last of an endemic group of South
American mammals. Initially described as a rhinoceros-sized rodent,
it is now thought to be a distant relative of the rhinoceros itself.

is today the largest, while at the same time peculiar to the continent on which
Toxodon was discovered'. This idea appealed to Darwin, for while he and Owen
could both subscribe to the law of succession of types (the replacement of species
by similar forms on each continent), it took on a deeper, evolutionary significance
for Darwin. Thus, even after Owen had transferred *Toxodon* to the Pachydermata,
Darwin continued to stress its rodent credentials, referring to it as 'intimately
related to the gnawers'. The teeth are certainly striking in this respect: their long,
narrow, curved shape; the absence of roots so the tooth could grow continuously
from the base as it wore down at the top; and in the incisors, the band of enamel
only at the front, producing a sharp, chisel-like gnawing edge. Although these
features can be found in other species, they are particularly prevalent among the
rodents. The idea of rhinoceros-sized rodents evidently tickled Darwin's fancy as
well – he wondered to his sister Caroline what tremendous cats must have preyed
on them in those far-off days.

A further intriguing question was how *Toxodon* might have lived. On the basis
of the relatively small space for its brain, Owen's first conclusion was that this

was an animal of limited intelligence. More significantly, he drew attention to features of the skull that he considered indicated a semi-aquatic or amphibious habit. The bony cups that had held the eyeballs extended vertically beyond the roof of the skull, and the nostrils appeared to be pointing upwards, both of which suggested to Owen behaviour similar to that of the hippopotamus, living much of its life partially submerged in water. The great forward-pointing incisors, he surmised, would have been used to uproot plants growing by the riverside. This idea was initially accepted not only by Darwin but by many palaeontologists of the time.

Fuller knowledge of the skeleton and habitat of *Toxodon* has revealed no special affinity for water. The very high-crowned molars and wide mouth are those of an open-country grazer, and analysis of carbon isotopes within the bone mineral suggests a grass-dominated diet, at least in the Pampas region where Darwin's fossils were preserved. The animal weighed in at around 1–1.5 tonnes, and anatomical study of the knee joint reveals a remarkable locking mechanism showing that *Toxodon* could stand for very long periods while feeding.

Toxodon is now known to be a member of a much larger group of South American mammals extending back some 60 million years but now wholly extinct. *Toxodon platensis* was the last surviving member of the group, and probably the largest. Owen himself first suggested the name Toxodontia in 1853, giving it the same status as other major groups of hoofed mammals and underlining its distinctive nature. Later, other related South American fossils were added, to form the Notoungulata ('southern ungulates'), a grouping still considered valid. Its members ranged from animals resembling a large rabbit (*Paedotherium*, see pp.74 and 75), to larger quadrupeds with bony 'horns' on their face, to the giant *Toxodon* itself.

The first discovery of any member of this group was in fact made by Darwin's fossil-hunting rival (and later collaborator) Alcide d'Orbigny, collecting shortly before Darwin in an area just to the north of the latter's hunting-ground at Bajada de Santa Fe (now Paraná, Argentina). By the time he came to describe it, Owen had already coined the name *Toxodon*, so d'Orbigny named his species *Toxodon paranensis*, and indicated that it came from earlier beds than Darwin's finds. These conclusions have been confirmed by later research – the species is now known as *Dinotoxodon paranensis* and is recognized as being of Miocene age, some eight million years old.

'A gigantic llama!'

Within six weeks of Darwin's purchase of the *Toxodon* skull he was 1,500 miles (2,500 km) to the south, in Patagonia near the tip of South America. Here, at the natural harbour of Port St Julian, he unearthed remains hardly less strange or significant. At Port St Julian, above a bed of gravel that Darwin had traced for hundreds of miles along the coast, a reddish muddy deposit was found near the surface; it had probably accumulated in the slow-flowing lower reaches of a river. At several points the mud filled deeper channels, and in one of these Darwin discovered a group of massive, beautifully-preserved bones. It was clear that an entire skeleton had originally lain there, and much of it remained. Darwin noted in his *Geological Diary* that several of the back vertebrae were lined up in a chain and joined to the pelvis, while 'nearly all the bones of one of the limbs, even to the smallest bones of the foot, were embedded in their proper relative positions'. This indicated immediately that the carcass, when buried, had still been united by flesh or ligaments, so could not have been very long dead or the bones would have dispersed.

Darwin gathered up every bone he could find. He speculated in letters home that the remains might be those of a mastodon or elephant, but admitted that he had no real idea what they were. Although his anatomical knowledge was limited, his caution with the St Julian find was prescient because the bones turned out to represent another highly distinctive group of extinct mammals that was completely unknown to science at the time.

By January 1837 Owen had formed some preliminary ideas about Darwin's fossil mammals, and regarding the skeleton from St Julian wrote to Lyell on 23rd of that month excitedly describing it as 'a Gigantic Llama!' The three

Darwin's sketch of the geology at Port St Julian, in southern Patagonia. The skeleton later named as *Macrauchenia* was enclosed in the dark infill (centre left) of the uppermost layer A.

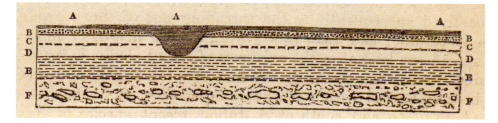

of them were in close contact and Darwin was party to this conclusion. Then, on 17 February, with Darwin in attendance, Lyell gave his Presidential address to the Geological Society of London. He praised Darwin's work in South America, mentioning the discovery of the 'gigantic llama', as well as the armadillo-like carapace and the 'rodent… the size of a rhinoceros' (*Toxodon*). These finds, he explained, suggested a general law of similarity between the extinct and living inhabitants of each continent, which had previously been limited to the finding of fossil kangaroos in Australia. Darwin had also been much struck by the apparent connection; in one of his very first private notes on evolution, written shortly afterwards, he specifically invoked the link between the 'extinct Guanaco' and its living relatives in speculating whether 'one species does change into another' (see Chapter 6).

The guanaco and vicuña, together with their domesticated forms the llama and alpaca, are small members of the camel family that live only in South America. In his initial assessment of the St Julian fossils, Owen had focused on the neck vertebrae (see p.89), striking in that they were longer than any known to him apart from those of a giraffe. However, it was not their length that suggested a connection to the camel family, but a detailed anatomical feature. In almost all mammals, the neck vertebrae have a hole on each side through which threads an artery carrying blood to the head. The exception is the camels, where the artery shares a passage with the spinal cord for most of its course through the vertebra. Owen found the condition in the St Julian fossil to be precisely like that of the camels. This was the basis for his initial assessment that had so much impressed Lyell and Darwin. But when he came to study the fossils in more detail, Owen was forced to change his mind.

From the time of Cuvier, the major division among the hoofed mammals had been based not on the structure of the neck but of the feet. The group to which the camels belong, which Owen later termed artiodactyls ('even-toed'), place the main weight of their body on two equally enlarged middle toes. In most of them, including the camels, the upper bones of these toes become fused into a single element. Darwin had collected most of the toe bones from one of the feet of the St Julian animal – and when Owen fitted them together the foot was nothing

The guanaco, *Lama guanicoe* – the South American cousin of the camel, parent species of the domestic llama, and initially considered a living relative of *Macrauchenia*.

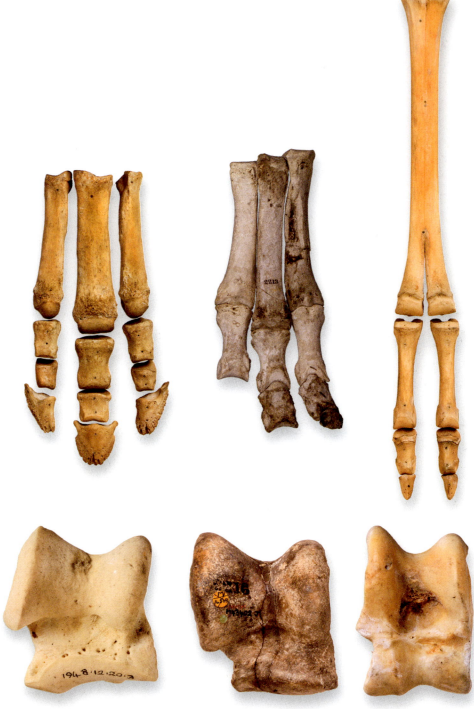

like that of a camel – or any other artiodactyl. There were three toes, of similar size, the central one being the largest, and all of them separate. Such a foot was like that of a rhino or a tapir – members of the group now termed perissodactyls ('odd-toed').

This conclusion was cemented by the structure of another bone – the ankle bone or astragalus. This, Owen enthused, would have been the bone of choice for any anatomist if he could choose only one for identification, and in this case had been 'most fortunately secured by Mr Darwin'. The astragalus is a small bone that connects the hind leg with the foot. In artiodactyls, including camels, it is shaped like a pulley at the top and bottom, allowing both the leg and foot bones to rotate around it. In other mammals, only the top of the bone has a rotation surface. The St Julian astragalus was nothing like that of an artiodactyl, and again resembled most closely the tapir as well as *Palaeotherium*, an ancient fossil mammal today recognized as an early perissodactyl.

In consequence, when Owen wrote up the St Julian animal for his monograph on the *Beagle* fossil mammals, he placed it in the group then containing the perissodactyls, not with the artiodactyls. But he emphasized that the importance of the animal lay in its forging a link between the two groups, which 'cannot but be viewed with extreme interest by the zoologist engaged in the study of the natural affinities of the animal kingdom'.

Owen wrote to Darwin in December 1837, revealing the name he had chosen for the creature – *Macrauchenia patachonica*. The genus name was formed from the Greek words macros (long) and auchen (neck), but also echoed the scientific name then in use for the llama – *Auchenia*. The species name refered to Patagonia

OPPOSITE TOP Toe bones of a tapir (left), Darwin's *Macrauchenia* skeleton (centre), and a guanaco (right). The three, fully separate, toes of the tapir and guanaco contrast with the two toes and fused upper bone of the guanaco. The *Macrauchenia* foot is 12 in (30 cm) long; the others are not to scale.

OPPOSITE BOTTOM The ankle-bone (astragalus) of a tapir (left), Darwin's *Macrauchenia* skeleton (centre), and a guanaco (right). The tapir and *Macrauchenia* are similar but the guanaco has an extra 'pulley' shape at the bottom. The *Macrauchenia* bone is 3 in (8 cm) long; the others are not to scale.

where it had been found. Darwin replied to Owen saying he thought the name was a very good one – adding that he was sorry to hear Owen had been unwell and hoped it was not because of the pressure he had been placing on him to finish his *Macrauchenia* text!

Owen later made a further brilliant deduction regarding *Macrauchenia* that has remained unnoticed. It appears that the St Julian skeleton was not the only fossil of *Macrauchenia* discovered by Darwin. In his 1845 monograph on teeth, Owen described a molar of an unknown animal found among Darwin's collection from Punta Alta. No part of the skull or dentition had been found with the St Julian skeleton, so no direct comparison was possible. However, in 1845, the British Museum purchased the fossil jaw of an unknown animal from Pedro de Angelis, an antiquary and traveller in South America. Owen recognized the similarity of its teeth to the Punta Alta molar, and both of them to teeth of the tapir, rhinoceros and the extinct *Palaeotherium*. Since the skeleton of *Macrauchenia* shared features with those forms, and since 'no other pachyderm of the size required to correspond with that indicated by the jaw and teeth … has hitherto been discovered in South America', he deemed it highly probable that the teeth belonged to the same genus. Darwin's tooth has unfortunately not survived, but the lower jaw remains in the Natural History Museum, London collection and later finds prove that it belonged to *Macrauchenia* as Owen surmised – and by extension so did the Punta Alta molar.

Owen's impression of the animal, from the parts at his disposal, was accurate enough – massive (we now know it weighed around a tonne), long-legged and with its neck held erect like a guanaco or llama, not flexed like a camel. When a skull of *Macrauchenia* was later discovered, it was found to be horse-like in general shape but with the great peculiarity of having a huge nasal opening on top of the skull between the eyes. By analogy with living species like elephants and manatees, this probably indicates that the animal possessed a short trunk, although some have suggested it simply housed muscles to tightly close the nostrils during dust storms. In its locomotion, the anatomy of the limb bones suggests that *Macrauchenia* was particularly adept at swerving and dodging – a

Neck vertebra of Darwin's *Macrauchenia* skeleton, shown in side and under-views. Its elongated shape indicates a long neck, but not the relationship to guanacos that Owen initially surmised. The length of the vertebra is 7 in (18 cm).

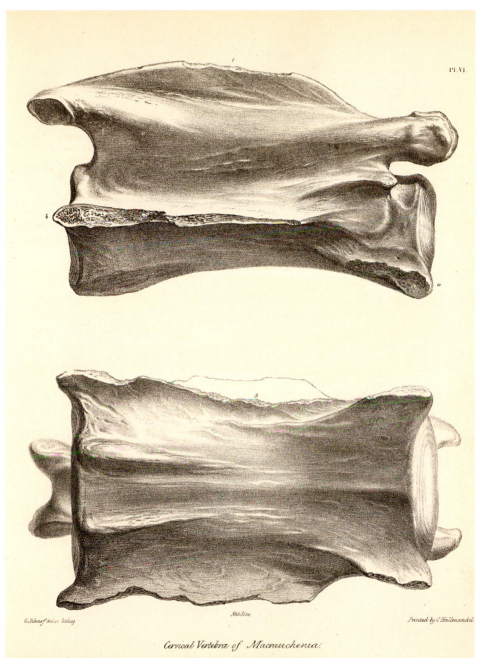

PL.VI.

G. Scharf del et lithog.   Nat.Size.   Printed by C.Hullmandel.

Cervical Vertebræ of Macrauchenia.

Published by Smith, Elder & C°. 65, Cornhill, London.

strategy that would have served it well in avoiding its main predator, the sabre-toothed cat *Smilodon*. *Macrauchenia patachonica* had a wide range, from Patagonia to Peru, and its habitat was probably woodland or savannah, as the form of its teeth and the carbon isotopes analyzed from its bones suggest leaf browsing, rather than grass eating, as its main source of food. With its long legs and neck it may have been a high browser like the giraffe, but instead of a long tongue it had its proboscis to gather food, somewhat like an elephant.

As far as its relationships are concerned, we now know that *Macrauchenia* had nothing whatever to do with camels. The characters of the feet that define the artiodactyls, and that *Macrauchenia* so conspicuously lacks, have stood the test of time, and the similarities of the neck vertebrae are the result of convergence – in other words, they had evolved independently in the two groups.

Just like *Toxodon*, later finds showed *Macrauchenia* to be the last-surviving member of a larger group, named the litopterns by the Argentinian palaeontologist Florentino Ameghino in 1889.

Macrauchenia patachonica, last of a uniquely South American group of mammals. The long neck and trunk were adapted for browsing on trees.

The litopterns arose some 60 million years ago, and the group spanned a wide range of sizes down to 88 lb (40 kg) or less – the size of a small deer. Only the huge *Macrauchenia* and its close relatives show the peculiar nasal opening and presumed trunk, a feature that developed in the group starting around 20 million years ago.

Both the litopterns and the notoungulates (*Toxodon* and its allies) diversified during the long period when South America was an island, in 'splendid isolation' as palaeontologist George Simpson aptly described it. Each arose from within the group of primitive, ungulate-like mammals called condylarths that were dispersed across the Old and New Worlds 65–35 million years ago. The relationship of the two groups to living groups of mammals, however, has been shrouded in mystery. Various proposals have been made, but the issue was not resolved until 2015 when fragments of collagen protein were extracted from well-preserved *Toxodon* and *Macrauchenia* bones. The protein sequences suggest, first, that litopterns and notoungulates are each other's closest relatives – a remarkable finding considering how different they are in body form. Second, the two share similarities with living horses, rhinos and tapirs (the perissodactyls, or odd-toed ungulates), making it very likely that they and the perissodactyls diverged from the same or closely related condylarth ancestors. The story has come full-circle, for when Richard Owen first defined the 'Toxodontia' in 1853, he suggested that it was related to precisely this group, on the basis of common features in the teeth and skull. Similarly for *Macrauchenia* he noted perissodactyl-like features in the foot skeleton, and Ameghino, in defining the Litopterna (the formal name for the litopterns, including *Macrauchenia*), considered it to be related to the Perissodactyla. Many anatomists since have favoured other features and different relationships, but the protein data warrant a fresh look at Owen's and Ameghino's 19th-century observations.

CHAPTER 3

❧

Petrified forests

INDIVIDUAL PLANT FOSSILS USUALLY comprise only part of the original organism, and Darwin's *Beagle* collections are no exception. The finds consist of either wood, or leaves, or compressed peat known as lignite. His collection is relatively modest, but his field observations on fossil plants included one of the most spectacular discoveries of the entire voyage.

Fossil wood

In many places on his travels, Darwin found petrified wood, either within the sediments in which they had originally been buried, or loose, having been washed out of their source deposit. In September 1833, for example, on the banks of the Paraná River in northern Argentina, close to where he discovered remains of mastodon and other megafauna, Darwin came across large blocks of petrified wood, and collected samples. He recognized that these came from a bed below that containing the mammals, and were significantly older – the wood was accompanied by sharks' teeth and the shells of extinct oysters. These deposits are now known to be middle Miocene age, some 15–10 million years old.

If wood becomes buried, one of three things may happen. It may disintegrate and not be preserved; it may become compressed into lignite or coal; or it may,

Leaves of *Glossopteris*, a seed-fern from the coal deposits of New South Wales, Australia. Darwin found similar leaves, some 300–250 million years old, when visiting the region in 1836.

Silicified wood, 3½ in (9 cm) long, of Tertiary, probably Middle Miocene, age (15–10 million years ago), collected by Darwin in the Paraná River area, Argentina.

over a long period of time, become petrified. Darwin described the wood from the Paraná as silicified – a form of petrification where the fossil is composed of silica (or quartz), the mineral familiar as the main component of sand. Subsequent microscopic examination of one of Darwin's specimens from this locality revealed that it is actually composed of calcite (calcium carbonate). Both of these minerals commonly contribute to the petrification of wood.

The commonest process leading to petrification is called permineralization. Wood contains many long, hollow conduits that transport water and nutrients up the trunk or along a branch. These are made of living cells, the contents of which disappear after death, leaving a space within the rigid cell wall. If the wood is buried in sediment through which water is percolating, minute quantities of silica or other substances dissolved in the water may crystallize out in the spaces within the wood, eventually filling them. This creates a replica of the original – technically a cast – that can be extremely detailed down to the microscopic level. The time taken for the process depends greatly on the conditions of burial and the composition of the percolating water. In ideal circumstances it can happen in as little as 50,000 years, though it may often take millions. Sometimes a further, slower process takes place, whereby the hard tissues of the wood – cellulose and lignin – are themselves replaced by minerals, so that the entire original structure of the wood is preserved in great detail.

Darwin noted the 'firm and universal' belief in the areas he was visiting that rivers had the ability to instantly convert any substance into stone – a belief doubtless fuelled by local people finding petrified wood. In December 1834, he discovered a fossil trunk in an ancient sandstone deposit on the Pacific coast of Chile. The wood was more fully petrified in some parts than others, which led him to realize that the process had taken place after burial and through slow percolation of silica-rich seawater through the sediment. He concluded, 'This observation [is] most important as proof of general facts of petrified wood, for here the inhabitants firmly believe the process is now going on'.

Different forms of fossilized wood are given Latin names ending in *-xylon*. A given form can often be identified only to the broad group of plants from which it came, but sometimes a more precise identification is possible. This generally requires the wood to be cut to produce clean sections in several planes to see its internal structure. Cut across a branch or trunk, the wood of coniferous trees is relatively uniform, with many small conducting cells called tracheids. The rays – lines radiating from the centre to the edge of the trunk like the spokes of a wheel – are relatively narrow. In the wood of most flowering plants, by contrast, the water is conducted by wider cells called vessels, and the rays are generally wider than in conifers. Details of the size, density and arrangement of the various structures can sometimes allow a more precise identification of the group or even species from which the sample came.

Although expert identification of his specimens awaited his return to England, Darwin was clearly aware of these basic distinctions (probably as a result of his botanical training from Henslow), for in both his *Geological Diary* written during the voyage, and in his letters home, he made broad identifications in many cases. He described the wood from the Paraná, for example, as dicotyledonous – the group that most flowering trees belong to (the main exception being the palms). Darwin recognized the significance of this in suggesting a lower limit for the age of the fossils and the deposit from which they came, for while the conifers and their allies have a much more ancient fossil history, now known to extend back at least 300 million years, flowering plants started to become common only in the middle of the Cretaceous Period, some 100 million years ago. Darwin considered most of his finds of fossil wood to date from the Tertiary Period, and in this he is now considered broadly correct, placing the remains younger than 66 million years old.

In one of their most arduous but rewarding excursions, a party of 25 men, Darwin and FitzRoy included, made their way up the Santa Cruz River in three small boats over three weeks in April to

A Darwin specimen of fossil wood from Chiloé in Chile, rediscovered in 2012 at the British Geological Survey. This thin section for microscopic examination was made by Robert Brown, a pioneer in the use of the method to identify fossil plants.

May 1834 (see map p.19). En route, Darwin found 'quantities of coniferous and ordinary dicotyledonous wood' scattered on the surface. Three months later, after the *Beagle* had crossed to the Pacific coast of Chile, he described seeing in a valley near Santiago an 'immense quantity of petrified wood… enough to fill [a] cart'. Then at Lacuy in the northwest of Chiloé Island (see map p.19), Darwin collected at one spot specimens illustrating the variety of ways wood could be preserved. The sandstone contained specimens that were silicified, others that were black and carbonaceous, and one where the wood had been converted to pyrites.

TOP Fossil wood, 3½ (9 cm) long, from the Tertiary of the River Santa Cruz, Argentina. Many of these specimens were sampled by Darwin from much larger trunks or branches that he encountered.

ABOVE Fossil wood, 5 in (12 cm) long, collected by Darwin from the Quaternary (the last 2.6 million years), of Iquique, Chile. The specimen shows naturally compressed wood, possibly at the point where a branch diverged.

Finally, he was gratified with more substantial finds. On the small island of Lemuy in the Chiloé group, Darwin found on the beach 'many large fragments of silicified wood', and was 'much pleased at last to find the silicified wood in situ; a large trunk, thicker than my body, throwing off branches, stood out of the sandstone… in the very position it was silicified'. He added that 'the form and structure of the wood is beautifully distinct – transparent quartz filling up the vessels'.

But the best was yet to come. During his return trek across the Andes in March 1835, Darwin took a pass through the Uspallata range, a small chain of mountains to the east of the main Cordillera. Noting that the rocks included sedimentary deposits of volcanic origin, he began to search for silicified wood, which he had found in similar deposits on the shores of the Pacific. 'I was', he wrote, 'gratified in a very extraordinary manner'. On 1 April, in the area known as Agua de la Zorra, he chanced upon a petrified forest of some 50 trees, standing almost upright (see map p. 143). The next day he wrote an account of the find in his notebook, then expanded on it in his *Geological Diary*, and shortly after, in exuberant terms, described his discovery in letters to Henslow and to his family. From these we gain a vivid account of the scene.

The stumps were all between 1–1½ ft (30–45 cm) in diameter, varying in height with the tallest of them around 6½ ft (2 m) from the ground, just over Darwin's head. Many stood within an area of around 160 ft (50 m) across, with a few outlying ones up to 500 ft (150 m) away. Some were within a metre of each

A massive log of petrified wood in present-day Patagonia, similar to those encountered by Darwin in Chile.

other, and Darwin's imagination was stirred: 'I saw the spot where a cluster of fine trees had once waved their branches on the shores of the Atlantic'... 'If they again could possess leaves and branches they would form an elegant cluster in an open country'. His invocation of the Atlantic, even though he was much closer to the Pacific Ocean, reflects his breadth of vision in time and space, since he was on the eastern flank of the Andes and the trees long pre-dated, he believed, the emergence of Patagonia from beneath the ocean (see Chapter 4). Importantly, he noted that all the trunks were inclined at an angle of around 20–30 degrees from the vertical, while the strata in which they were rooted were inclined by a similar amount from the horizontal, making it very likely that they were preserved in the ground in which they had grown. Two short pieces, 'as thick as a man's arm', perhaps representing broken branches, were seen lying horizontally.

The trees were not all identically preserved, however. Eleven of them were silicified, their quartz infill preserving the original concentric lines of the wood, and when a piece of sediment was removed from near the base of the trunk, it preserved a mould of the original pattern of the bark, 'circularly furrowed with irregular lines'. The remaining trees, by contrast, were formed from calcium carbonate and preserved no such detail, so that Darwin knew they were trees only from their identical position and form to the silicified ones. These evoked another image, this time a biblical one: 'The cylindrical columns of snow white, coarsely crystallized carb: of lime were very conspicuous & reminded me of Lot's wife turned into a pillar of salt.'

The identification of the wood had to await Darwin's return to England. For this (and for identification of other fossil plants from the voyage) Darwin sought

The base of a tree-trunk, some 16 in (40 cm) in diameter, exposed today at Agua de la Zorra in Chile. Darwin was searching for the source of fossil wood fragments when he came across 50 petrified standing trees at the site.

Specimen of silicified fossil wood, 4 in (10 cm) in length, brought home by Darwin from the Agua de la Zorra forest.

the help of Robert Brown, Keeper of Botany at the British Museum and noted microscopist. Darwin had met him shortly before the *Beagle* voyage, when he had sought advice on what kind of microscope he should take with him. Brown was evidently impressed by the finds, for Darwin wrote to his friend Leonard Jenyns: 'Tell Henslow, I think my silicified wood has unflinted Mr. Brown's heart, for he was very gracious to me.'

Brown told Darwin that the wood 'partakes of the character of the Araucarian tribe (to which the common South Chilian pine belongs)'. The 'Chilean pine' is the tree now popularly known as the monkey-puzzle tree. It belongs to an ancient family of tall coniferous trees, the Araucareaceae, originally distributed worldwide but now restricted to South America, Australasia and parts of Southeast Asia. The petrified wood from Agua de la Zorra is now named *Agathoxylon*, and while it may belong to the monkey-puzzle family it could also pertain to other extinct groups of conifers or seed-ferns where the wood is similar. In the 1990s, new fieldwork was undertaken at the site by Argentinian palaeontologists. They reconstructed the original forest as comprising the *Agathoxylon* trees to a height of some 15–65 ft (16–20 m), but towering above them to 65–80 ft (20–24 m) a second type of tree, belonging to the extinct group known as seed-ferns – not true ferns but flowerless seed-plants with fern-like leaves. The researchers also found evidence of roots at the bases of some of the trees, and a thin layer of fossilized soil in which they had grown. The rocks also revealed impressions of the leaves of true ferns, representing the understorey of the forest.

Microscopic image of piece of petrified wood, *Agathoxylon pararaucana*, from Darwin's Forest. The exquisite preservation is evident – the image is only 0.8 mm wide. Three growth rings are shown, each comprising uniform small water-conducting cells (tracheids).

The aspect of this fossil forest that intrigued Darwin the most, however, was the mode of its formation and what that implied for the history of the Andes. The thick deposit of sandstone in which the trees were 'planted' is a sedimentary deposit that must have formed in a deep lake or under the sea; Darwin favoured the latter. He correctly reasoned that as the trees must have grown on dry land, the area had been uplifted prior to the development of the forest. A conundrum arose, however, because the deposits surrounding and above the trunks appeared to Darwin also to be of sedimentary origin, so he was forced to assume a subsequent subsidence of the land to a great depth below sea level, as the deposits extended hundreds of metres above the trees. Finally, the land must have risen once again to form the mountain chain, elevating the buried trees to their present position. Darwin realized what an extraordinary claim this was, but it was a logical deduction given his interpretation of all the sediments as laid down by water.

A more satisfactory solution was demonstrated only in 2004, when detailed study of the sediments around the trees showed that they were not the result of slow underwater accumulation but, quite to the contrary, sudden burial by the outflow of a volcanic eruption. Darwin himself had observed that the material surrounding the trees was 'mudstone including broken crystals and particles of rock'. This material is now known to represent a pyroclastic flow, erupted material that flows down the slopes of a volcano at high speed, engulfing everything in its path. This is the process responsible for the burial and preservation of people and animals at the Roman city of Pompeii in AD 79, but the phenomenon was quite unknown in Darwin's time. The sudden burial of the living trees also explains how the deposit preserved detailed impressions of their bark.

Darwin believed the rocks and the trees to be of Tertiary age, but they are now known to be very much older – Middle Triassic, around 235 million years ago. Darwin correctly recognized that the forest had grown before the Andean mountains had been uplifted, but while the latter is now known to have been a relatively recent event by geological standards (around 30–15 million years ago), Darwin imagined it to be much more recent still, post-dating even the arrival of people in South America. His timing may have been out, but his understanding of the growth of the forest was not. All of his fossil plant finds were, for him, primarily indicators of past landscapes. He noted that while coastal Chile is today clothed with forest in the south, further north it is a desert (the Atacama) as a result of the rain shadow of the Andes. The coast to the west of the Uspallata range, where he had found his fossil forest, is today dry and treeless; before the mountain uplift it had evidently been moist and forested.

The petrified forest was the discovery that, in addition to the giant mammals, began to make Darwin's name even before he returned to England. Darwin had written to Henslow to describe the find on 18 April 1835. Henslow was so impressed that on 16 November he read an abstract of Darwin's account at a meeting of the Cambridge Philosophical Society, and included it in a pamphlet for distribution to the members. Two days later Sedgwick read an account to a meeting of the Geological Society in London (see Chapter 6). As for the site itself, it is no longer possible to see it as Darwin witnessed it; souvenir hunters over the years have reduced most of the trunks to stumps. In 1959, however, on the centenary of the publication of *The Origin of Species*, a plaque was erected to mark the discovery. The area is now a nature reserve, the 'Parque Paleontológico Araucarias de Darwin'.

A Gondwanan forest of the Mesozoic, similar to that discovered by Darwin at Agua de la Zorra. *Araucaria*-like trees grow in the distance, while ferns and horsetails line the river bank. To the right are coniferous trees and an overhanging ginkgo.

The plaque marking the site of Darwin's Forest at Agua de la Zorra, Chile, replacing the original 1959 plaque and erected to mark the bicentenary of Darwin's birth.

Seventeen days after his return, Darwin set out on a ride of some 420 miles (675 km) taking eight weeks, northward from Valparaíso to the Chilean city of Copiapó (see map p.19). He soon encountered a 'wonderful number of huge embedded logs of silicified wood' and doubted 'if any part of the world can produce such immense quantities of silicified, blackened, often metalliferous wood as here lies scattered'. The abundance continued – in the valley of Copiapó he recorded large quantities embedded in sandstone and thousands of huge blocks scattered about on the surface. One trunk must have belonged to a tree 20 ft (6 m) in circumference – 'what an extraordinary prop!!' exclaimed Darwin; 'The spectacle of these trees of stone would astonish anyone.' Based on associated shells, he estimated their age as Early Cretaceous, some 140–120 million years ago by today's reckoning. He recalled with approval Humboldt's suggestion that the very hard silicified wood remained on the surface after the much softer rocks in which they were embedded had eroded away. In the *Journal of Researches* published after his return to England he continued to marvel 'how surprising it is that every atom of the woody matter in the great cylinder should have been removed and replaced by silex so perfectly, that each vessel and pore is preserved!' We now know that rocks derived from volcanic matter are particularly effective in the petrification of wood due to their mineral content and their porous nature that allows water to percolate.

Ancient trees of the Antipodes

After her long voyage across the Pacific, the *Beagle* landed briefly in New Zealand and then at Sydney, Australia, on 12 January 1836. In Darwin's collection of fossil woods are specimens (some of them sectioned by Brown) from Illawara in New South Wales. This locality, near the town of Wollongong, was not on Darwin's itinerary during his brief excursion from Sydney, but his notebook entry indicates that he was told of large quantities of silicified wood being found there, and it is likely he was given some specimens. Preserved in the collections of the Natural History Museum, London, they have been identified, since Brown's time, as *Dadoxylon*, a form of fossil wood now considered identical to *Agathoxylon*. They therefore probably represent the monkey-puzzle family that today and in recent geological history is characteristic of South America and Australasia. These specimens are much older than most of the fossil wood collected by Darwin himself, of Tertiary age, and older even than the Triassic forest of Agua de la Zorra. The deposits from which the Illawara wood very likely came are Carboniferous to Permian, some 300 million years ago.

BELOW This Darwin sample from Illawara is a naturally polished cross-section of a branch or trunk showing well-defined annual growth rings. It is 6 in (15 cm) wide.

ABOVE Sample of ancient silicified wood in the Darwin collection from Illawara, New South Wales, showing well-defined growth rings. It is 2 in (5 cm) wide. The specimen is probably from a species of the monkey-puzzle family.

Fossil wood from Tasmania similar to, and probably related to, the cypress. The Darwin specimen, 3½ in (9 cm) long, shows the fibrous texture of the wood.

From Sydney the *Beagle* travelled to Tasmania (then known as Van Diemen's Land) where, during a 10-day stay, Darwin again acquired specimens of fossil wood, apparently from the central part of the island and again possibly given to, rather than found by him. One specimen has been identified as *Cupressinoxylon*, in other words a conifer with wood similar to, and probably in the same family as, the cypress family. The age of the material is uncertain but is probably of Cenozoic age.

Returning to the mainland in March 1836, the *Beagle* anchored for a week at King George Sound in the southwestern corner of Australia, preparing for her voyage across the Indian Ocean. Darwin, as always, had a scientific objective, and walked with Fitzroy to the end of a narrow promontory called Bald Head to witness a site mentioned by many previous navigators - a profusion of upright, petrified branches that some had thought were trees, others corals. Darwin reports that after careful study, he and FitzRoy arrived at the same conclusion. The structures were calcareous, made of a substance similar to that which forms stalagmites, and that had led some to consider them corals. Darwin and Fitzroy, however, were in no doubt that they were the remains of trees, from the roots branching out at their bases and their fibrous, woody structure that could be seen in places. The stumps emerged from sandstone containing shells of land snails, and the companions concluded that the trees must have been buried in dunes formed by wind-blown sand that was then hardened by percolating, calcium-rich water. The wood itself had decayed, leaving spaces that were gradually filled in by crystals of calcium carbonate. Subsequently the softer sandstone eroded away, leaving the casts of the trees standing proud. It was a brilliant piece of geological deduction.

Imperfect coal

As well as petrified wood, Darwin encountered deposits of lignite at various points on his travels. Lignite forms when partially decayed plant matter – peat – becomes buried, and under conditions of increased temperature and pressure loses moisture and becomes compacted. Mined for fuel in many parts of the world, it is sometimes known as brown coal; Darwin himself referred to it as 'imperfect coal'. Geologically, it is an intermediate stage in the formation of true, or bituminous, coal. Lignite may be formed from varying proportions of wood, leaves or other parts of plants; encountering beds of lignite in Chiloé, southern Chile, Darwin wondered which was predominant, recognizing the significance of the question for understanding the nature of coal. He took a boat to the nearby island of Lemuy 'anxious to examine a reported coal mine', and there observed 'extensive horizontal layers of black lignite, structure of wood yet very visible: is said to communicate much heat in furnace'. Further up the Chilean coast near Concepción he visited another lignite mine, but noted that, 'The mine is not worked, for the coal when placed in a heap has the singular property of spontaneously igniting', as a result of which several ships had apparently caught fire. The spontaneous combustion of lignite and coal is a process that is now well known but still not fully understood. It is thought to be due to the combination of oxygen in the air with constituents of the coal in a reaction that gives off heat, causing loss of moisture, and the further propensity of the coal to heat up until a threshold is reached and the pile catches fire.

Lignite ('imperfect coal') collected by Darwin at a mine on the island of Lemuy, Chile.

In New Zealand over Christmas 1835, Darwin was given a sample of lignite from a deposit on the west coast used as fuel by local people. The specimen is still in the collection of the Sedgwick Museum in Cambridge. 'The vegetable fibres', noted Darwin, 'are so very distinct that the whole substance almost resembles an altered peat'. This led him to ponder a question debated among geologists at the time – were deposits of similar composition necessarily of the same age? If the New Zealand lignites should prove to be of Tertiary age, much younger than the coal measures of Europe, it would demonstrate that coal formation was 'owing, not to the age, but to the circumstances under which it accumulated'.

Only once on the voyage did Darwin witness a deposit of true (bituminous) coal. Landing at Sydney, Australia, on 12 January 1836, he set out for a 10-day excursion into the interior. At the head of the Wolgan Valley he encountered a layer of black coal some 1 ft (30 cm) thick, and recognized that it belonged to the same formation that was extensively mined north of Sydney in the area of Newcastle, named after the coal-rich town in the home country. These coals are Permian in age (some 300–250 million years old).

Leaves of the southern continents

Although wood and lignite were by far the most abundant remains of plants seen by Darwin, he collected a few fossil leaves of types that, much later, became highly significant for our understanding of the Earth's history. At the bay of St Sebastian on the Argentinian side of Tierra del Fuego, at the base of a high cliff, the rock was laminated, and splitting apart the layers Darwin found the impressions of 'a great number of the leaves of trees; I believe they are from the beech which now so abounds on the mountain'. Alongside them were fossil shells, barnacles and crabs indicating a marine environment, so Darwin imagined in past times 'a bay where muddy sand and leaves … brought down by brooks, might collect [and] be deposited'.

The leaves were later examined by Darwin's friend the botanist Joseph Hooker, who considered them to belong to species of deciduous beech, differing from the living beech which, as Darwin had noted, formed the great proportion of trees in the local forests. Darwin used the botanical name *Fagus* to describe the finds; this is the genus of beech with which he was familiar in England. There are a dozen or so living species across Europe, Asia and North America. At the time, the beeches of the Southern Hemisphere were placed in the same genus, but in 1851 the

Partial leaf of the beech *Nothofagus* of Tertiary age, collected by
Darwin in Tierra del Fuego. The leaf impression, some 1 in (3 cm)
long, comprises a thin film of carbon remaining from the original leaf.

Betula antarctica. Sydney Parkinson pinxt 1769

LEFT A watercolour by Sydney Parkinson of the leaves of the living southern beech, *Nothofagus*, from Tierra del Fuego – a specimen collected in 1769 by Sir Joseph Banks during Captain Cook's first voyage.

BELOW A typical strap-like leaf of the seed-fern *Glossopteris* from the Permian (300–250 million years ago) of New South Wales, now recognized as a marker for the supercontinent Gondwana.

German-Dutch botanist Karl Ludwig von Blume recognized them as belonging to a separate genus, *Nothofagus*. This grows today in southern South America, Australia, New Zealand, New Guinea and a few nearby islands. Its fossil distribution is similar, with the addition of Antarctica.

When Darwin was examining the coal deposits at Wolgan in Australia, he made a second discovery of fossil leaves that perfectly complements the first. The find is not recorded in any of his contemporary notebooks or diaries, but is mentioned in his book *Volcanic Islands*, published in 1844 that also contained 'some brief notices of the geology of Australia'. There he reports that within shale deposits alternating with the coal seams, he found 'leaves of the *Glossopteris brownii*, a fern which so frequently accompanies the coal of Australia'. There are no specimens of this kind in the remaining Darwin collection, so they may have been lost, or it may be that he simply described his field observations to Brown or Henslow on his return. The identification is quite plausible, however, for fossils of *Glossopteris* are so common in the Wolgan area that hikers are invited to keep an eye out

DRIFTING CONTINENTS

In the first half of the 20th century, *Glossopteris*, and to a lesser extent *Nothofagus*, played a critical part in the recognition that the continents have moved around the Earth's surface over long periods of time and were formerly joined in differing combinations. This is the process of 'continental drift' now explained by the theory of plate tectonics. Specimens of *Glossopteris* are found in rocks of Permian age (around 300–250 million years old), with a distribution that matches the positioning of the southern continents when the 'jigsaw' of their outlines at that time is pieced together as shown in the map. Gondwana is now thought to have formed some 500 million years ago; it later joined with the northern continents to form Pangaea but then separated again. It gradually broke up between 180–80 million years ago, carrying its cargo of 'Gondwanan' fossils on each of its constituent continents.

The southern beech *Nothofagus*, that Darwin collected in Tierra del Fuego, has a similar fossil distribution (though excluding Africa) and has been considered also to be a Gondwanan relict. Unlike *Glossopteris* it is still extant, and its fossil history is considerably shallower, dating back only to around 80 million years ago. This, and the similarity of DNA among *Nothofagus* from different continents, has led some researchers to suggest that part of its modern distribution formed after the break-up of Gondwana, so that the beeches on New Zealand, for example, arrived by dispersal across open ocean.

The supercontinent Gondwana in Permian times, showing areas where fossils of the seed-fern *Glossopteris* (orange) have been found. (1) South America, (2) Africa, (3) Madagascar, (4) India, (5) Antarctica, (6) Australia.

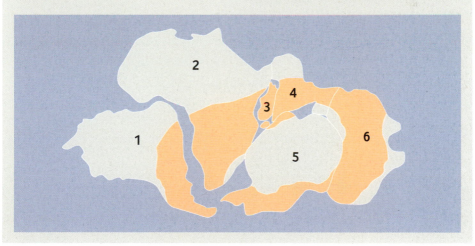

for them, and the long, strap-like fronds are characteristic (*Glossopteris* means 'tongue-fern'). *Glossopteris* was not a true fern, however, but a shrub- or tree-like member of the extinct seed-fern group described earlier.

A final, evocative 'remnant of a lost vegetation', as Darwin described it, was discovered during one of his excursions in Tasmania in February 1836. In a quarry behind Hobart town, limestone was being extracted, probably for processing in a nearby kiln to produce lime for use in building materials. Limestone (calcium carbonate) forms under various circumstances, but in this case Darwin described it as travertine – a form of limestone deposited on land by calcium-rich mineral springs. Darwin found the Hobart travertine to 'abound with distinct impressions of leaves', and his list of fossil plants indicates that he brought home six specimens. These were examined by Robert Brown at the British Museum, who reported that there were four or five distinct types represented by the leaves, none of them belonging to living species. 'The most remarkable leaf', noted Darwin, 'is palmate, like that of a fan-palm, and no plant having leaves of this structure has hitherto been discovered in Van Diemen's Land [Tasmania].' The deposits, at Geilston Bay, are now recognized as Late Oligocene in age (around 30–25 million years ago), but it is uncertain whether any of the Natural History Museum's existing specimens from this site are from Darwin's collection.

The constant juxtaposition of Darwin's fossil discoveries with the living biota that surrounded him is vividly evident in this instance. Immediately after his description of the Hobart fossil flora in *Journal of Researches*, he describes his climb up nearby Mount Wellington, where 'in many parts the gum trees grew to a great size, and the whole composed a noble forest… Tree-ferns flourished in an extraordinary manner… forming so many most elegant parasols'. Darwin saw his fossils not just as geological specimens but as witnesses to a past world as rich and varied as that of today but, crucially, different.

OPPOSITE ABOVE AND BELOW Could this be a Darwin specimen? The 12 in (30 cm) block of travertine, from Geilston Bay, Tasmania where Darwin collected, was discovered in the collection formerly curated by Robert Brown, who had studied Darwin's fossil plants.

CHAPTER 4

~

Marine life

BY FAR THE MOST ABUNDANT and widespread fossils encountered by Darwin on the *Beagle* voyage were the shells and other hard parts of invertebrate animals, most of them of marine origin. His collections were dominated by molluscs of many species, but also included the remains of sea urchins, sea lilies, brachiopods, crabs, barnacles, bryozoans and corals. While many of these were named as new species by the specialists appointed by Darwin – principally because so little geological collecting had been done in the areas he visited – this was not their primary interest for him. Just as for geologists today, fossils such as these were indicators of past habitats, markers of geological time, and tools for ordering and correlating the deposits in which they occurred.

From the *Beagle's* first landfall, at St Jago in the Cape Verde islands on 16 January 1832, Darwin recorded the rocks that he saw and the fossils that he found in them, collecting specimens as he went. He would do so at every subsequent stop on the journey, up to and including the Azores nearly five years later as the ship headed for home. St Jago is the remnant of a large, extinct volcano, and Darwin observed along much of its southern coast (and on the small offshore Quail Island) two bodies of volcanic lava with a layer of white, sandy rock sandwiched between them. This band of limestone held an abundance of fossils, and Darwin collected

A 28 to 16 million year old giant oyster, *Crassostrea patagonica*, collected by Darwin at Port St Julian. The shell, 8½ in (22 cm) long, is shown close to life size; some individuals grew to 12 in (30 cm).

Collected by Darwin at St Jago in the Cape Verde Islands, these spherical fossils, now known as rhodoliths, are the calcareous skeletons of free-living red algae. They range from 1–2 in (2–5 cm) in diameter.

the remains of limpets, oysters, snails and sea urchins, as well as spherical bodies that he later recognized as the calcareous skeletons of red algae. The age of this layer and its enclosed fossils was uncertain for Darwin, but on his return home the British conchologist George Sowerby identified 18 species of mollusc, three of which he considered extinct. The deposits are now believed to have formed some three-quarters of a million to a million years ago, in the Quaternary Period.

In the course of the voyage Darwin collected fossils spanning a vast range of geological time, from a few thousand years to some 400 million years ago. The fossils from different periods were all of intense interest to him, for different reasons. The most ancient remains provided evidence of some of the oldest life then known, while those of intermediate age recorded changes in the Earth's biota through time and marked major events such as the formation of the Andes. We begin at the most recent end of the scale, where the shells and other fossils that

Darwin collected from relatively superficial deposits, at many locations around the South American coast, were of crucial importance in two major respects. First, they provided evidence of the environment and age of the mammal fossils he had collected, in many cases at the same localities (see Chapter 2). Second, they bore witness to major recent changes in the level of the South American continent, which Darwin was the first to document.

The elevation of South America

One of Darwin's major achievements on the *Beagle* voyage was his discovery that much the southern half of the South American continent had been uplifted in relatively recent geological times. A pivotal role in this conclusion was played by his findings of fossil shells. At many places he visited, from the estuary of the Plata in the north, to Tierra del Fuego in the south, Darwin found sea shells on plains elevated far above the present beach level or the reach of the highest tides. The shells lay on the surface or were loosely embedded in the soil or sandy deposits, and were often present in large numbers.

Most of the mollusc fossils found by Darwin were either gastropods (with a single, usually snail-like coiled shell), or bivalves (clams, with a hinged double shell). Marine molluscs are easily distinguished from those of terrestrial or freshwater origin; their shells are generally thicker, and their forms are identifiable as species or families that live exclusively in salt or brackish water, such as oysters and true limpets. When found well above the reach of modern tides, their remains must indicate that either the sea level had fallen, or the land had risen up. Darwin followed Lyell in assuming the latter, since even at a continental scale the magnitude of displacement would be much less than an equivalent fall in the level of the sea, which would have to have been global since all the oceans are interconnected. Also, he had noted at St Jago that the level of the uplifted marine beds varied along the coast, proving that it was 'not by subsidence of water'.

Darwin encountered the shell-beds at various elevations above the present sea level. The altitudes of the different plains were measured by the *Beagle's* officers as part of their surveying work, and Darwin noted these carefully. Sometimes he measured them himself using a barometer. Above St Joseph's Bay on the Patagonian coast, for example, 100 ft (30 m) above sea level, sandy hillocks abounded with mussel (*Aulacomya*), limpet (*Nacella*) and spindle shells (that Darwin named *Fusus*), as well as large barnacles. Further south at Port Desire,

Rock containing brackish-water bivalves, *Erodona mactroides*, found near the Rio Plata but above its tidal range, demonstrating recent uplift. This specimen was given to Darwin by Sir Woodbine Parish.

mussel and limpet shells were again numerous, this time on plains at 250 ft (75 m) and 330 ft (100 m). Still further south, at Port St Julian, there were abundant shells on a 90 ft (27 m) plain, and Darwin was struck by the similarity in height to that of St Joseph's, some 470 miles (760 km) distant, suggesting that they were formed during the same episode of upheaval.

In May 1834, the crew's boat trip up the Santa Cruz River in Patagonia provided Darwin with his best opportunity to observe the effects of uplift, for as the party ascended the river valley, successively higher plains came into view. At a point 105 miles (170 km) from the Atlantic and at a height of around 300 ft (90 m), he found broken shells of marine snails in the bed of the river. Further up the valley, 140 miles (225 km) from the ocean and at a height of 440 ft (140 m), he collected worn shells of limpets and whelks. The finding of marine fossils far inland convinced Darwin that there must previously have been a wide inlet of the sea, followed by uplift of the beds it had deposited.

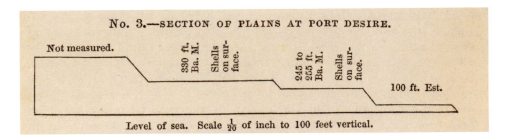

'Section of Plains at Port Desire', from Darwin's *Geological Observations on South America*. One of several diagrams to show a staircase of plains strewn with marine shells, demonstrating sequential uplift of the land.

Not only did the shell-beds provide evidence of uplift, but it seemed clear that this had happened in very recent geological times. Even in the field, Darwin noted that the shells, at all levels of elevation, were the same species as those still living on local coasts, and this was confirmed when the collections were identified by Sowerby and the French naturalist Alcide d'Orbigny. Even the relative proportions of species were similar to those of a shell collection made on the modern beach. Furthermore, the ancient shells frequently retained part of their original colouring – the mussel shells almost always held a blue tinge, and other species such as barnacles sometimes their pinkish hue. Darwin placed these shells in a flame and found that they emitted an animal odour – not only had the mineral component of the shell remained, as in more ancient fossils, but also some of the organic component that we would now identify as shell proteins. Altogether, Darwin wrote in his *Geological Diary*, 'This is to me a certain proof, that these plains have been uplifted from the bottom of the ocean within a recent period.'

Before leaving the Atlantic coast of South America, Darwin collected his thoughts in an essay entitled *The Elevation of Patagonia*. He had observed uplifted beds of marine shells along some 1,200 miles (1,900 km) of coastline and they probably, he surmised, extended to 1,600 miles (2,600 km) or more. In this he was challenging Lyell, who saw elevation as something far more localized. 'If on some future day', Darwin concluded, 'I shall be able to prove that the West coast has been elevated within the same period, it will almost render it certain that the whole S. part of continent has been elevated.'

Proof was soon at hand. Two months later, on the island of Chiloé on the Pacific coast of Chile, Darwin was guided by Mr Williams, a fellow Briton and

Captain of the port, to 'an immense bed' of shells on a table-land, which he measured by barometer as some 350 ft (106 m) above the level of the sea. In other parts of the island, 20–30 ft (6–10 m) above high-water mark, the land was thickly coated by venus clams (*Ameghinomya*) and mussel shells (*Mytilus*), 'the species now most abundant on this line of coast'.

Further north, on the small island of Quiriquina in the bay of Concepción, Darwin and the *Beagle's* assistant surgeon Mr Kent went fossil-hunting, finding beds of barnacles and various species of marine mollusc at heights of 20, 164, 400 and 625 ft (6, 50, 122 and 190 m). Darwin noted how the shells were more decayed the higher the level, conforming to the idea that the highest levels were the oldest, having been subject to successive periods of uplift. In the area around Valparaíso, Darwin found the remains of mollusc shells, sea urchins and barnacles as high as 1,300 ft (400 m), including conical top-shells (that he named *Trochus*) 'with colour quite perfect'. Then, riding north from Valparaíso to Copiapó in April and May 1835, he noted 'immense quantities' of shells on plains at 50-80 ft (15–24 m) and at 200 ft (60 m). They were marine shells, chiefly rock snails (*Concholepas*), venus clams and the bivalve *Mesodema*, and Darwin noted that they were in valleys some tens of miles from the present coast, showing that the sea had once penetrated well inland.

Darwin was aware of one factor that might undermine his interpretation of shell-beds as evidence for uplift. Shellfish of various kinds were a staple part of the diet of local people in coastal areas and, as Darwin put it, on Chiloé 'the inhabitants carry immense numbers of these shells inland… to considerable heights and distances from the sea'. To assure himself that some, at least, of the raised shell-beds were of natural origin, he examined all the localities carefully. Generally, it was the extent and continuity of the shell-beds that persuaded Darwin of their natural origin; he had noticed in Tierra del Fuego that shells collected by people were patchy and formed heaps that persisted long after they had been abandoned. By contrast, the shell-bed at San Lorenzo on the coast of Peru, for example, was about a mile long and 150 ft (45 m) wide. At two points near Valparaíso shell-beds were found 'on the edge of a precipitous cliff 200 ft high', with no obvious route down, and no fresh water nearby.; 'Why should people bring shell-fish to such places?' asked Darwin. The shell-beds also generally contained various species mingled together, whereas piles accumulated by human activity tended to be of a single kind. Finally, many of the limpets, mussels and other species were of very small, young individuals, some of the shells barely a quarter of an inch (6 mm) across. Darwin took a local fisherman to see the remains, and he 'ridiculed the notion of such small shells having been brought up for food'.

BALANUS (BALANUS) LAEVIS,
var. COQUIMBENSIS, G. B. Sowerby

PLEISTOCENE. INLAND FROM HERRADURA BAY,
COQUIMBO, CHILE.

Holotype of *Balanus coquimbensis*, G. B. Sowerby, in
Darwin, 1846, Geol. Observ. on S. America, appendix,
p. 264, pl. ii, fig. 7; figd. as *Balanus laevis*, var. *coquimbensis*
G. B. Sowerby, by Darwin, 1854, Ray Soc. Monogr.
Balanidae, p. 227, pl. iv, fig. 2a.

Presd. by Charles Darwin, 1858.   [38436]

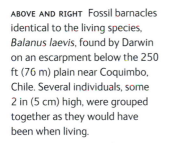

ABOVE AND RIGHT Fossil barnacles identical to the living species, *Balanus laevis*, found by Darwin on an escarpment below the 250 ft (76 m) plain near Coquimbo, Chile. Several individuals, some 2 in (5 cm) high, were grouped together as they would have been when living.

The origin of the staircase-like series of gently sloping plains on which the shells were strewn still required an explanation. Darwin was certain that the ancient shells had been empty and lying in shallow water or on an ancient beach when it was uplifted. The species he had collected were always those which today live close to shore and are thrown up onto the beach by the tide. The fossil molluscs were, moreover, often encrusted with barnacles, and perforated with the characteristic small holes produced by various marine creatures. On the island of San Lorenzo off the coast of Peru, Darwin made the particular observation that upraised rock snails (*Concholepas*) had barnacles and the cases of marine worms (serpulids) adhering to their *inner* surfaces – they had 'clearly lain untenanted at the bottom of the sea'.

While in the field, Darwin saw in the succession of raised terraces evidence for the 'theory of numerous successive elevations, which Dr Lyell has so strongly supported'. Imagine a beach sloping into the sea and at its head a cliff. Now imagine a sudden uplift of the whole coastline by 100 ft (30 m) or more. The old cliff would now form an inland escarpment, and at its base a long plain (the former beach and adjacent seabed) sloping down to the sea. Over time, the sea would eat into the sloping plain to form a new cliff. This formed a second step of the staircase, and at the next sudden uplift the process would repeat. In favour of this model Darwin cited historical examples of uplift. It was well known in scientific circles that the coast of Chile had risen within living memory, and at Lacuy (Chiloé), Mr Williams told Darwin that 'during the last four years, the sea has sunk or land risen about four feet'. 'In the above period', noted Darwin, 'Chiloé has suffered from a severe shock of an earthquake.' Captain FitzRoy described the effects of the Concepción earthquake of 1835 that the *Beagle* crew had witnessed (see p.142): 'The southern end of the island of St. Mary was uplifted eight feet, the central part nine, and the northern end ten feet… Great beds of mussels, patellæ, and chitons still adhering to the rocks were upraised above high-water mark; and some acres of a rocky flat, which was formerly always covered by the sea, were left standing dry, and exhaled an offensive smell, from the many attached and putrefying shells.' What more compelling image for the formation of raised shell-beds could there be?

Later, Darwin came to realize that uplifts of 100 ft (30 m) or more were unlikely to have been sudden. Instead, the land had risen in a gradual movement, albeit aided by the 'small sudden starts, such as those accompanying recent earthquakes'. The staircase of terraces would then be explained by significant pauses in the process of uplift, during which the sea would erode the coast to

The flat surface of a marine terrace (raised beach) some 490 ft (150 m) above sea level north of Port Desire. Lower (younger) steps in the terrace sequence have been eroded by the action of the sea in this area.

produce a new line of cliffs. Darwin was led to prefer this explanation partly from a realization that all the shells on the raised terraces were those that live close to the shore; had sudden elevations of 100 ft (30 m) or more taken place, deeper-water species would have been exposed and preserved as well.

When the *Beagle* reached the South Atlantic for the second time in July 1836, there was an interesting coda to Darwin's interpretation of high-elevation shell-beds. On the volcanic island of St Helena he found accumulations of mollusc shells 20 ft (6 m) below the surface but several hundred feet above sea level. A local resident, Robert Seale, presented him with a parcel of shells from another location, some 1,700 ft (520 m) above the sea. Similar species were seen in the upper levels of a limestone quarry, where bones and even eggs of birds were also preserved. The molluscs, Darwin reported, had been assumed to be marine in origin, and to indicate recent elevation of the island from the sea. This would have accorded with his interpretation of shell-beds in many other places, but here he thought differently. First, he was convinced the shells were those of land snails, so there was no need to invoke elevation to explain their presence on a hillside. Second, Darwin was much struck by the unique flora of the island, which he called a 'little world within itself'. This degree of endemism argued for a high antiquity of the island – a distinctly evolutionary perspective since it implied that divergent species required longer to develop. By the same token, the island's emergence could hardly have been recent.

On Darwin's return home, George Sowerby confirmed that all the shells from St Helena were of terrestrial species, and moreover, six out of seven were extinct. Darwin had already concluded this for the commonest and most conspicuous species (which he identified as an awl snail, *Bulimus*), as local residents told him they had never seen one alive. The explanation for their recent demise, Darwin suggested, was the loss of woodland that had covered much of the island as recently as the 18th century, coupled with the impact of introduced pigs and goats. These conclusions have all been borne out by recent research. The native plants and molluscs of St Helena are around 80% endemic (species found nowhere else on Earth). However, 22 species of mollusc have become extinct since 1600, including six species of *Chilonopsis* that include Darwin's '*Bulimus*', the losses being attributed to habitat degradation from plant and animal species introduced by humans.

Molluscs and megafauna

Punta Alta, in the bay of Bahia Blanca some 310 miles (500 km) south of Buenos Aires, was the location of many of Darwin's most important finds of fossil mammals, including the remains of *Megatherium*, *Mylodon* and *Toxodon*

No. 15.—SECTION OF BEDS WITH RECENT SHELLS AND EXTINCT MAM-MIFERS, AT PUNTA ALTA IN BAHIA BLANCA.

Darwin's sketch of the cliff at Punta Alta. Mammal bones and mollusc shells were found mainly in the lower gravel (A), but also in red mudstone of the 'Pampean Formation' (B) and overlying gravel (C).

LEFT AND BELOW Fossil marine
shells collected by Darwin from
the 'Pampean Formation' at Punta
Alta: left, *Tegula patagonica*; 4 in
(10 cm) long; below, volute shells,
*Zidona dufresne*i, ½–¾ in
(1–2 cm) in diameter.

RIGHT Fossil coral, *Astraea*,
collected by Darwin from the
'Pampean Formation' at Punta Alta.
The coral, 2¾ in (7 cm) long, is
the hard exoskeleton that housed
small, anemone-like polyps.

(see Chapter 2). The gravel deposits in which the bones were found were also rich
in fossil shells, the most abundant being slipper limpets (*Crepidula*), top snails
(*Tegula*) and venus clams (*Amiantis*); others included oysters, mussels, and rare
scallops (*Aequipecten*). As well as molluscs, Darwin found pieces of the shells of
barnacles, and fragments of the hard skeletons of a coral (*Astraea*).

Encrusting some of the fossils were the lace-like remains of bryozoans (known colloquially as moss animals or sea mats). These tiny colonial organisms had, since the 17th century, been grouped with corals and sponges as 'Zoophytes' and considered intermediate between plants and animals. Darwin found living examples, too, and early in the voyage grouped them simply as corals, but by 1834 he had begun to study them in detail, and recognized that they were not only animals but unrelated to true corals. He referred the leaf-like colonies to the bryozoan genus *Flustra*, with which he was familiar from his work on the Scottish coast with his Edinburgh tutor Robert Grant.

As with the shells strewn over the plains, Darwin noted that shells in the underlying deposits 'appear to me to be exactly the same species which now exist on the beach, and it is to be especially remarked the proportional numbers of each species are about the same'. This was a clear indication that the fossil bones associated with them must be of very recent geological age. He was alert to the possibility that some of the bones might in theory have been washed out of earlier deposits and redeposited with the modern shells, but some at least were clearly contemporaneous. This was illustrated by the complete skeleton of the ground sloth *Scelidotherium* (see p.49), which 'must have been washed entire to near the place where found, and therefore a very modern animal coeval with present shells', so emphasizing its 'extreme lateness of existence'.

Even stronger evidence came to light at Port St Julian in Patagonia, where Darwin unearthed the giant, llama-like animal later named *Macrauchenia* by Richard Owen (see p.83). The bones lay close to the top of the cliff, just below a land surface only 90 ft (27 m) above sea level. Behind this was a further plain at 350 ft (106 m), strewn with the shell species commonest on the modern coast, so even this was relatively recent, and by the logic of uplift the *Macrauchenia* must have been later still. Moreover, the bones themselves were mingled with still-extant marine species, some of the shells retaining their original colour.

OPPOSITE TOP A living encrusting bryozoan (sea mat) showing feeding zooids, each around 0.5 mm in diameter, their tentacle crowns protruding through pores in the communal skeleton.

OPPOSITE BOTTOM Fossil bryozoan skeleton collected by Darwin from the 'Pampean Formation' at Punta Alta; width of image 1 in (2.5 cm). This is an encrusting species of the family Electridae; each pore is the opening of an individual zooid.

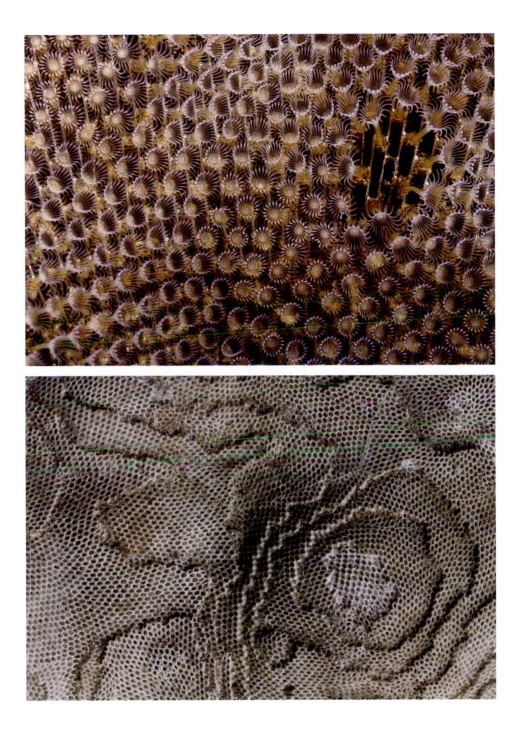

Marine shells, many of them retaining original colour, preserved in sandy sediments some 6,000 years old, 15 ft (4.5 m) above present sea level in Santa Cruz province, Argentina.

The association of extinct mammals with extant mollusc species had further theoretical importance for Darwin. As he put it, 'We here have a strong confirmation of the remarkable law so often insisted on by Mr. Lyell, namely, that the longevity of the species in the Mammalia, is upon the whole inferior to that of the Testacea [sea shells].' Species endured for different periods of time and, moreover, their extinctions were not all synchronous – the molluscs had persisted while the large mammals had died out. Similar patterns were seen in the fossil record of Europe and North America, noted Darwin, and were a boost for Lyell against catastrophists such as d'Orbigny who envisaged periodic cataclysms when almost all life died out at once, to be replaced by new creations (see Chapter 6). Hence d'Orbigny had suggested that all the Pampean mammalian fossils had been washed out of earlier beds and later redeposited along with the recent shells, a concept rejected by Darwin.

Under the microscope

Not content with fossils evident to the naked eye, Darwin wanted to know about microscopic ones, too. In Berlin, Christian Ehrenberg was considered the supreme authority on microbial life, naming thousands of species and interesting himself not only in living forms but fossil ones as well. With Joseph Hooker acting as intermediary, Darwin sent Ehrenberg samples of sediment scraped off fossil bones from Punta Alta, and Ehrenberg reported finding two species of 'Polygastrica' –

single-celled, microscopic organisms now known as diatoms. Both of them, quoted Darwin approvingly, were 'decidedly marine forms'. Ehrenberg also listed a number of tiny structures as 'Phytolitharia'; the nature of these was unknown at the time but they are now known not to be microscopic organisms in their own right but tiny silica particles found in plant tissues. Several of those identified for Darwin have since been recognized as deriving from terrestrial grasses.

The beds at Punta Alta also included bands of hard, white, calcareous deposit that Darwin called by its local name, tosca. Again Darwin sought the services of a leading specialist to examine it – William Carpenter, a medical doctor, neurologist and marine biologist. Darwin reported that Dr Carpenter 'can perceive distinct traces of shells, corals, Polythalamia and rare spongoids'. 'Polythalamia' were single-celled organisms with multi-chambered shells, now termed Foraminifera. These studies were pioneering in attempting to reconstruct the ancient environments of larger fossils by detailed examination of their enclosing sediments.

Finally, at the Paraná River near Santa Fe, Darwin had found two mastodon skeletons protruding from the cliff, but had been able to recover only fragments of teeth (see p.64). He nonetheless scraped the red mud from one of them and sent it to Professor Ehrenberg, who found it contained mostly freshwater diatoms. The difference from the other localities, all on the coast, was interesting, as Santa Fe is currently some 217 miles (350 km) from the sea; the bones had likely been preserved in deposits of the Paraná River itself.

Bringing the story up to date

Darwin's conclusions on the uplift of southern South America have stood the test of time, as has, in part, his model for the formation of successive raised beaches, now also known as marine terraces. Unknown to Darwin and his contemporaries, however, was the effect of recent ice ages, alternating roughly every hundred thousand years between glacial and interglacial periods, and with profound effects on the global sea level. During each of the glacial periods, vastly expanded ice sheets at high latitudes locked away so much of the world's water that sea levels fell by 330 ft (100 m) or more. At this time, many currently offshore areas were dry land. In the intervening, relatively brief, warm interglacials, sea levels rose and these areas were flooded again. Marine terraces are now understood to have formed by the interplay of this sea level oscillation with ongoing regional uplift. During low-stands, terrestrial and freshwater deposits accumulated on coastal plains. At high-stands, the sea ate into

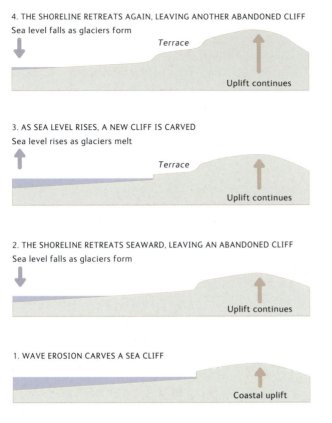

4. THE SHORELINE RETREATS AGAIN, LEAVING ANOTHER ABANDONED CLIFF

Sea level falls as glaciers form

Terrace

Uplift continues

3. AS SEA LEVEL RISES, A NEW CLIFF IS CARVED

Sea level rises as glaciers melt

Terrace

Uplift continues

2. THE SHORELINE RETREATS SEAWARD, LEAVING AN ABANDONED CLIFF

Sea level falls as glaciers form

Uplift continues

1. WAVE EROSION CARVES A SEA CLIFF

Coastal uplift

Stages, from bottom to top, in the formation of a marine terrace (raised beach). With further uplift and sea level change, a second terrace would be formed, then a third, and so on.

these deposits, creating a beach and carving out a cliff. With continuing uplift, each interglacial beach formed a new step in a staircase of marine terraces. In southern South America, uplift has been gradual, just as Darwin believed, but instead of periodic interruptions to its progress to explain the sequence of terraces, we now invoke the rise and fall of sea level.

Thanks to modern chronological methods, we can now date the formation of the marine terraces on the Patagonian coast. The lowest terrace corresponds to Darwin's 20–30 ft (6–10 m) plain atop the fossiliferous deposits at Punta Alta. This terrace, from radiocarbon dating of the fossil shells, dates to the last sea-level high stand, earlier in the present interglacial about 8,000–6,000 years ago. The next widespread terrace, at 50–65 ft (16–20 m), dates to the last interglacial about 120,000 years ago. At around 27 m, Darwin's 90-foot terrace is dated to the previous interglacial, around 200,000 years ago; a 130–165 ft (40–50 m) terrace to 300,000 years ago, and so on. During this period the average rate of uplift can be calculated as around a tenth of a millimetre per year. Some of the highest terraces, at 500–650 ft (150–200 m), may be more than two million years old.

The sediments underlying the marine terrace surfaces are mostly red silt with layers of hard tosca rock. Darwin named these deposits the Pampean Formation, and he was certain that they must have formed in the estuaries of large rivers like the modern Plata. We now know that deposits of this type are for the most part terrestrial. The fine, uniform particles are typical of wind-blown sediment that settles on land and may then be picked up and redeposited elsewhere by rivers. The tosca is also a terrestrial deposit, now known as calcrete, and is formed from calcium carbonate that has leached from the soil. Gravel layers, such as those seen by Darwin at Punta Alta, were laid down by rivers at a time when the coastline, due to glacial sea-level fall, was far away.

How, then, are we to explain the presence of marine shells in these deposits, associated with the bones of land mammals? This was one of Darwin's main reasons for assuming deposition within an estuary, where land and sea meet. Most of the mammal fossils from Punta Alta were found in the lower gravel layer ('A' on p.122), and it now seems likely that the marine shells had been picked up by the river as it cut through the older marine terraces further upstream. The shells were then deposited along with the gravel and bones. Darwin's own observations support this idea, as he noted that the shells in the gravel layer were bleached and worn. The tiny fossils found by Ehrenberg and Carpenter had also been taken as evidence for marine deposition, but the study of such remains, now known as micropalaeontology, was then in its infancy and the range of forms pertaining to marine or freshwater environments was not well established. Darwin had finally noted that many of the bones from Punta Alta had the remains of marine animals attached to them: barnacles, bryozoans, and the tubes of marine worms (serpulids), potentially confirming marine deposition (see p.130). With characteristic candour he admitted, however, that 'I neglected to observe whether these might not have grown on them since being exposed to the present tidal action'; in the case of the skeleton of the ground sloth, *Scelidotherium*, found in a fallen block of sand on the beach, he was sure this was the case.

The position of the Punta Alta fossils, in deposits beneath the 6,000–8,000-year-old beach, but lower, in altitudinal terms, than the 120,000-year-old beach, places them between the present interglacial and the previous one. In other words they were contemporary with the last glaciation, which extended from around 115,000 to 12,000 years ago. Darwin was always cautious about assigning ages to fossil deposits, but based on the residual colour on some of the shells from the high plains

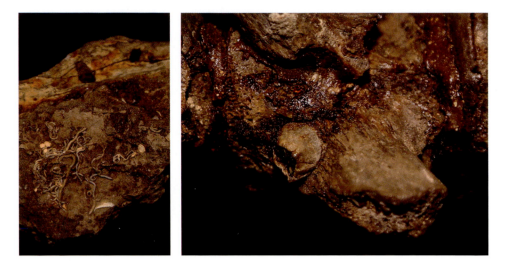

Modern marine animals growing on the skeleton of the ground
sloth *Scelidotherium* that Darwin found lying on the beach at
Punta Alta. Worm tubes on sediment attached to the ribs (left) and
a barnacle attached to a vertebra (right).

he suggested that 'within no great number of centuries all this country has been
beneath the sea'. His estimate was out by a factor of perhaps a thousand, and we
now know that in favourable circumstances shells can retain some of their colour
and protein (the source of the animal odour when burnt) for millions of years.

Shell-beds of the Tertiary Period

At the same time that Darwin was recording the Pampean deposits and their
spectacular mammalian fossils, he found fossil-rich beds of a very different nature
underneath them. Not only were these, from their position, clearly older, but the
shells they contained proved to be almost entirely of extinct species, in sharp
contrast to those found on top of the elevated plains and associated with the
mammals. Darwin correctly interpreted these beds as 'Tertiary' in age, equivalent
to all but the latest Cenozoic (see p.33).

The Tertiary sequence was generally similar wherever he looked – at its base, and
often exposed at the foot of cliffs, beds stuffed with giant oysters and other marine
shells. Above this, beds of varying thickness, generally pale in colour with fewer
fossils, and comprising limestone, gypsum and fine volcanic matter. At St Joseph's

Bay in April 1833, for example, Darwin described a lower bed comprising about a fifth by mass of shells, dominated by enormous oysters, themselves bearing traces of holes bored in ancient times by other molluscs and sponges. There were also the remains of scallops, barnacles and, in some places, masses of the calcareous skeletons of marine coralline algae. Encrusting the shells were various kinds of bryozoans, and specimens identified by Darwin as cup-corals. It was an unmistakably marine assemblage, and moreover, 'did not blacken or emit any bad odour under the blow-pipe'. In other words, all organic matter had disappeared, underlining a greater antiquity than the higher shell-beds.

RIGHT Scallop, *Zygochlamys actinodes*, collected by Darwin at St Joseph's Bay; 6 cm (2½ in) long. Predatory boreholes (probably by another mollusc) can be seen, while two juvenile oysters had settled on the right of the shell.

BELOW Beds of giant oysters on the coast of Patagonia today. Darwin collected from the 'great oyster formation' at various localities spanning hundreds of miles of coastline.

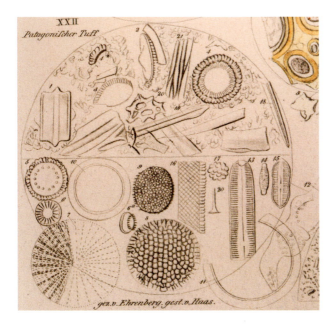

Ehrenberg's 1854 illustration of microfossils in Tertiary sediment collected by Darwin at Port Desire and Port St Julian. 1–16 are diatoms, 17–19 are phytoliths and 20–21 are inorganic volcanic particles.

By the time he had reached the southern port of St Julian in January 1834, Darwin was referring to the lower deposit as 'the great oyster formation'. Here the fossil deposits were 800 ft (245 m) thick and Darwin collected from several beds. One of the beds contained large quantities of sea-urchin shells – they were a flattened form that he identified as *Scutella*, now informally known as sand dollars. From the numerous barnacles Darwin deduced that the whole biota must have lived in relatively shallow seas, with the coast not too far away. He later sent the sea urchins to Alexander Agassiz, a noted expert, in Switzerland, and it was named as a new species. Another of Darwin's sea urchins, from Port Desire further to the north, was named *Monophora darwini* after its finder – it is now known as *Monophoraster darwini*.

The Miocene sand dollar, *Iheringiana patagonensis* – a sea urchin with a flattened form discovered by Darwin at Port St Julian. The shell is 3 in (8 cm) across but only ½ in (1 cm) deep.

At the top of the cliff at Port St Julian, the channel containing the *Macrauchenia* skeleton was cut directly into white, fine-grained rock that represented the upper part the Tertiary formation. After the voyage, Darwin submitted samples of this deposit to Professor Ehrenberg, together with similar material from Port Desire. Combining the two, Ehrenberg reported no fewer than 16 species of diatom, all of them thought to be marine. Fossil shells in this deposit were rare, but significant. Darwin noted that the species were the same or very similar to those of the lower oyster bed, implying that the entire 800 ft (245 m) of deposition had occurred during the same general epoch. A similar observation was made in sections at the Rio Santa Cruz, further to the south.

The northernmost point at which Darwin identified Tertiary marine deposits was near the town of Bajada de Santa Fe (now Paraná) in northeast Argentina, where he found sandy clays with 'vast numbers of large oysters' and other marine shells. The contrast with the overlying Pampean deposits, containing the bones

ABOVE Fossils collected by Darwin at Port St Julian. Left, gastropod, *Trophon sowerbyi*, 2½ in (7 cm) long – one of many new species named from Darwin's specimens, and right, barnacles, *Hesperibalanus varians*, the specimen is 3 in (8 cm) long.

RIGHT The large gastropod mollusc, *Adelomelon alta*, collected by Darwin from Tertiary deposits at Santa Cruz. The specimen is some 5 in (12 cm) long and includes fragments of other gastropod and bivalve species.

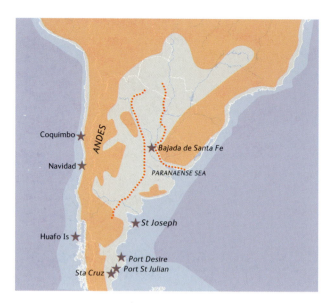

South America in the Middle Miocene (around 15–10 million years ago); land shown in orange. Stars indicate key localities where Darwin collected Tertiary marine fossils; many were far offshore at the time of their formation.

of large mammals and apparently freshwater diatoms (see p.127) was striking, especially since the locality was so far from the present sea – some 185 miles (300 km) from the Plata estuary and twice as far from the open ocean. Alcide d'Orbigny later identified from Darwin's collection a new species of scallop that he named *Pecten darwinianus* (now *Amusium darwinianum*).

When the *Beagle* crossed to the west coast of the continent, Darwin continued to find important deposits of Tertiary fossils, although they were patchier and did not appear to be outcrops of a single formation. Writing to Henslow on 28 October 1834, Darwin described fossils from south of Santiago and made a remarkable suggestion. Noting that the shells appeared to be more different from modern ones than those he had seen on the east coast, he added that 'it will be curious if an Eocene and Meiocene [sic] could be proved to exist in S. America as well as in Europe'. He was here invoking the brand new subdivision of the Tertiary Period proposed by Sir Charles Lyell in the third volume of his *Principles of Geology*. The book had been published in London in May 1833 and a copy had probably reached Darwin while he was in the Falkland Islands in March 1834. Lyell recognized three subdivisions of the Tertiary Period: the Eocene, Miocene and Pliocene, the latter further subdivided into two parts, and they could be recognized by the proportion of living mollusc species they contained. For the Newer Pliocene, over 90%; for the Older Pliocene, 30–55%; for the Miocene, around 20%; and for the Eocene, less than 5%. (Lyell later coined the term Pleistocene for the Newer Pliocene, the Older Pliocene becoming just Pliocene and the boundary between them adjusted from 90% to 70% recent shells.) Already by October 1834 Darwin seemed to be speculating that the Patagonian shells might be Miocene in age, and the small collection from Santiago might be from the Eocene. Darwin, half a world away, was probably one of the first (apart from Lyell himself) to employ the new scheme. As he himself pointed out, 'This rests on the supposition that species become extinct in same ratios over the whole world', since Lyell's work was based on European fossils. The giant oyster of Patagonia (see p.112), was a particularly clear example of an extinct species; local people told Darwin they had never seen one alive, and 'so remarkable a shell could not escape observation'.

Darwin collected Tertiary fossils at several other points along the Pacific coast of Chile, from Huafo Island south of Chiloé, to Navidad, 60 miles (100 km) from Valparaíso. Arriving at Navidad on 22 September 1834, he 'staid here the whole of the ensuing day, and although very unwell managed to collect mainly marine remains from beds of the Tertiary formation'.

A tusk shell, the scaphopod mollusc *Fissidentalium subgiganteum*, collected by Darwin at Navidad in Chile. The tubular shell is attached to sediment in which shells of other species are visible.

Evidence of marine life extended beyond shell-living invertebrates. In a valley to the north of Coquimbo, in a bed of mussels and oysters, Darwin found 'the teeth of a gigantic shark', likening it to *Megalodon*, known from the Tertiary deposits of Europe. This shark, now named *Otodus megalodon*, is recognized as one of the largest predators ever to evolve, with a length of up to 60 ft (18 m) and individual teeth 6 in (15 cm) or more in length. In the same deposit, moreover, were great numbers of large bones in a silicified state, far too heavy to collect but identified by Darwin, almost certainly correctly, as those of whales. He had previously found, in the hills above Port St Julian together with fossil oysters and scallops, 'a vertebra of some large, probably cetaceous animal' (i.e. a whale or dolphin); 'the diameter was about 7 inches & stone very compact & heavy'. The find is given a number – 1719 – indicating that it was collected, although the fate of the specimen is unknown.

At Navidad, and at two of his Argentinian localities, Darwin recorded further fish teeth in the Tertiary deposits, without further details. Some of these remains were evidently passed to Alexander Agassiz, who took them with him to the Museum of Comparative Zoology at Harvard, USA. Collected at Port St Julian, the specimens were identified only in 1945 as the teeth of a sawfish, a family of rays in which the teeth (actually modified scales) project from either side of an elongated rostrum.

A Miocene scene with the 60 ft (18 m) *Otodus megalodon*, a giant relative of the great white shark, in pursuit of the small toothed whale, *Squalodon*.

A living sawfish, *Pristis pectinata*. The saw is used to dig up shellfish and to slash at swimming prey.

Fossil 'teeth' of sawfish, *Pristis*, collected by Darwin at Port St Julian. The specimens, up to 3½ in (9 cm) long, are shown life-size.

Subsequent studies have shown that the 'Tertiary' assemblages collected by Darwin cover a range of ages. However, considering that the Tertiary had only just been subdivided in 1833, and that two further divisions were added later (the Paleocene before the Eocene and the Oligocene after it), Darwin's allocation of his fossil collections was broadly within range. Today, the deposits in southern Patagonia – at Port Desire, Port St Julian and the Rio Santa Cruz – are placed within the San Julian and Monte León formations, of Late Oligocene to Early Miocene age (deposited between 28 and 16 million years ago). The section described by Darwin at the mouth of the Santa Cruz River, in particular, is considered a landmark in our understanding of the geology of Patagonia.

The Tertiary deposits at Darwin's northern outpost at Bajada de Santa Fe are younger, and now define the Paraná Formation, dating to the middle part of the Miocene, around 15–10 million years ago. The marine deposits here, far inland, are key evidence for the last great incursion of the sea into the South American continent, forming a marine province that has been dubbed the Paranaense Sea (see map p.134). The fossils from St Joseph's Bay are of similar or younger age, around 12–5 million years old.

On the Chilean side, the Navidad Formation stands today as an important standard for the marine sediments and fossils of the mid-Tertiary in South America, although its precise age is the subject of debate. Some researchers consider the deposit to be quite late, in the region 10–4 million years ago, with fossils of earlier age (such as the shark's teeth) reworked from older deposits. Others prefer an Early to Middle Miocene age (23–12 million years ago) for the deposits and their fossils. The fossils from Huafo Island, near Chiloé in Chile, are of Pliocene age and therefore the youngest of all, some 5–3 million years old.

The Tertiary fossils provided Darwin with a striking illustration of another of Lyell's key principles. At several of the localities (St Julian, Santa Cruz, Huafo Island and Navidad), shallow-water marine species had been deposited through 800 ft (245 m) or more of sediment. This implied that the sea floor had subsided

A group of coiled gastropods, *Incatella chilensis*, each 1 in (2.5 cm) long, collected by Darwin in Chile, probably on Huafo Island.

by an equivalent amount during deposition; that, or the sea level had risen by an equivalent degree. Since the latter seemed implausible, there must have been considerable subsidence during the Tertiary Period which, given Darwin's evidence for later uplift, graphically illustrated the fluctuating nature of crustal movements.

Fossils from the age of dinosaurs

Below Darwin's Tertiary formations (largely Cenozoic in today's terms) lies the Mesozoic Era, separated from it by the most celebrated of geological events, the Late Cretaceous mass extinction that wiped out dinosaurs, most of the ammonites and many other groups. Darwin first discovered Mesozoic fossils on the second of the *Beagle*'s three expeditions to the southern tip of South America. In February 1834 the ship was anchored off the Brunswick Peninsula, across the Strait of Magellan from the islands of Tierra del Fuego. On 6 February, leaving the ship at 4 am, Darwin set out to climb Mount Tarn, the highest peak in the area at some 2,600 ft (800 m) above sea level. After a difficult climb through dense forest, he emerged into the open and at last attained the summit. 'The strong wind was so piercingly cold that it would

Mount Tarn in Tierra del Fuego, Chile, showing the stratified rocks where Darwin collected the first ammonite from South America, after a tortuous climb through the scrub below.

A Cretaceous marine scene, including the two-metre-long paperclip-shaped ammonite *Diplomoceras cylindraceum*.

Darwin's ammonite fossil, *Maorites*, from the summit of Mount Tarn – part of the outer whorl of the spiral shell. The specimen is of Late Cretaceous age and is 4 in (10 cm) long.

prevent much enjoyment under any circumstances', he wrote in his diary, but 'I had the good luck to find some shells in the rocks near the summit'. The finds were few but included fragments and impressions of two species of marine gastropod, a shell Darwin identified as a brachiopod but which got lost on the descent, and some pieces of echinoderm shell that he suggested as sea-urchin but were later identified as the stalked sea-lily, *Pentacrinites*. Most significant, however, were fragments of shells that Darwin described as 'a special Nautilus'. On another day, between torrential rain showers, Darwin explored the area around Port Famine, 9 miles (15 km) to the north of Mount Tarn, where the slate deposits were exposed close to beach level. The finds included bivalves, fragments of coral, and once again a shell that Darwin described as a nautilus.

Back home in England, all three of the mollusc specialists seem to have been involved in identifying the finds. D'Orbigny named the remains from Mount Tarn, Sowerby those from Port Famine, and Forbes added his own comments in a letter to Darwin. Darwin's 'nautiloids' turned out to be ammonites; both groups are cephalopod molluscs, relatives of modern octopus and squid, and most species have coiled outer shells. Nautiloids were commonest in the Paleozoic and then progressively declined, although six species survive today. Ammonites diversified in the Mesozoic but became extinct at the end of the Cretaceous or shortly after. In both nautiloids and ammonites, the living animal occupied the outermost, largest chamber of the shell, but a strand of tissue (the siphuncle) extended through the other chambers to regulate buoyancy. One way of separating the two groups is that the tube holding the siphuncle runs through the middle of the shell in nautiloids but round the outside in ammonites.

Darwin's specimens from Mount Tarn and Port Famine are believed to be the first ammonites ever recorded from South America. Moreover, his two specimens from Port Famine proved to be a remarkable species. Based on preserved portions some 2 in (6 cm) in diameter, Sowerby named it *Hamites eliator* but it is now identified as *Diplomoceras cylindraceum*. These large ammonites reached up to 6½ ft (2 m) in length, or 13 ft (4 m) around the coil of the shell, with three U-bends which have been described as forming a giant paperclip.

The ammonite from Mount Tarn was a quite different form, considered by d'Orbigny to be identical to a species he had named from Europe – *Ancyloceras simplex* (a partly coiled ammonite). However, recent examination of the specimen suggests that it is a slightly deformed fragment of a different genus, *Maorites*, that has a more typical spiral form and is restricted to the southern Gondwanan continents (see p.109). Ammonites are particularly valuable fossils for dating purposes, and both d'Orbigny and Forbes correctly allocated the material from Port Famine and Mount Tarn to the Cretaceous Period. Both ammonite species are now known to be Late Cretaceous forms, dating from some 72–66 million years ago, just before ammonites finally went extinct.

Four months later, on 10 June 1834, the *Beagle* passed through the Magellan Strait and entered the Pacific Ocean. There was to follow a remarkable series of events that deeply impressed upon Darwin the immense power of crustal movements of the Earth. Viewing the Andean mountains from the *Beagle* anchored off Chile, Darwin observed on 26 November 1834 that 'the volcano of Osorno was spouting out volumes of smoke'. On 19 January the *Beagle* was still in the vicinity and Orsono was erupting: 'At midnight the sentry observed something like a large star; from which state the bright spot gradually increased in size till about three o'clock, when a very magnificent spectacle was presented. By the aid of a glass, dark objects, in constant succession, were seen, in the midst of a great red glare of light, to be thrown upwards and to fall down again. The light was sufficient to cast on the water a long bright reflection.'

An even more dramatic event was to follow. At around 11 am on 20 February 1835 the city of Concepción was destroyed by one of the most severe earthquakes ever recorded in Chile. Both Darwin and FitzRoy, in their respective journals, gave graphic accounts of the event. They were onshore at the time, in the city of Valdivia, some 185 miles (300 km) to the south; even there, Darwin reported that 'the rocking was most sensible' as 'the world, the very emblem of all that is solid, moves beneath our feet'. In the ensuing weeks the *Beagle* was at sea and FitzRoy reported frequent aftershocks, likening them to an anchor jolting a ship when its chain runs out before reaching the bottom.

On 4 March the *Beagle* had anchored at Concepción and the following day Darwin and FitzRoy rode into the town. To Darwin, it was 'the most awful and interesting spectacle I ever beheld' – encapsulating the tension between the suffering and destruction they witnessed and their intense scientific interest in the earthquake itself. Barely a house was left standing, and the effects of the earthquake had been compounded by a massive wave 23 ft (7 m) higher than the highest natural tide – a tsunami. 'The force of the shock must have been immense', Darwin wrote to Henslow, for 'the ground is traversed by rents, the solid rocks shivered, solid buttresses 6–10 feet thick are broken into fragments like so much biscuit'. FitzRoy interviewed local

Locations where Darwin collected, or from where he was given, Mesozoic fossils. Also shown are the locations of the Agua de la Zorra Mesozoic forest, the Osorno volcano, and the Concepción earthquake.

residents and recorded in detail the raising of the land by up to 10 ft (3 m). For Darwin, having spent two years recording geological evidence for former uplift, the connection was obvious; he subscribed, with Lyell, to the view that the Earth's solid crust rests on a core of molten rock, and a build-up of pressure could lead to eruptions, earthquakes, or both. Eventually, even a vast chain of mountains like the Andes could have been thrown up by a combination of crustal uplift and the outpouring of volcanic lava.

Barely a fortnight later, Darwin was to see graphic fossil evidence of the power of uplift. Determined to study the geology of the Andes at close quarters, Darwin travelled to Santiago to prepare for a trek across the mountains. The expedition began with a three-day climb past massive beds of gypsum and volcanic rocks a mile in thickness. At the top of the Piuquenes pass, some 13,200 ft (4,000 m) above sea level, Darwin was surprised to see marine sedimentary rocks, and in a deposit of black slate he found, to his great joy, the remains of fossil shells. The walk had been arduous due to the rarefied air – Darwin reported that 'the exertion of walking was extreme, and the respiration became deep and laborious' – a condition known locally as the puna. But 'upon finding fossil shells on the highest ridge, I entirely forgot the puna in my delight. The inhabitants all recommend onions for the puna... for my part, I found nothing so good as the fossil shells!' Darwin was not the first to find the remains of marine organisms at high altitude, even in the Andes. But the power on the scientific mind of witnessing something with one's own eyes should not be underestimated. As Darwin himself later wrote: 'It is an old story, but not the less wonderful, to hear of shells, which formerly were crawling about at the bottom of the sea, being now elevated nearly fourteen thousand feet above its level.'

The Piuquenes Pass provided Darwin with another image that he would remember for the rest of his life. Turning to look back from the top of the pass, 'a glorious view was presented. The atmosphere so resplendently clear, the sky an intense blue, the profound valleys, the wild broken forms, the heaps of ruins piled up during the lapse of ages, the bright coloured rocks, contrasted with the quiet mountains of snow, together produced a scene I never could have imagined. ... I felt glad I was by myself, it was like watching a thunderstorm, or hearing in the full orchestra a chorus of the Messiah'. For Darwin in South America, like Humboldt before him, the scientific discovery of nature, and the aesthetic response to it, were intertwined.

The fossils on top of the Andes comprised abundant shells of the marine bivalve *Gryphaea*, other bivalves and gastropods, a probable brachiopod, and some ammonites – among them, species of massive size. One segment of ammonite shell, Darwin reported, 'was so large I could not carry it, being thicker than my arm although only a small segment of the curve'. A few specimens were collected, but as he wrote to Henslow shortly afterwards, it was late in the season with the danger of snow storms – 'I did not dare to delay, otherwise a grand harvest might have been reaped.' The importance of the shells was clear, however: 'I think an examination of these will give an approximate age to these mountains as compared to the strata of Europe.' The presence of ammonites indicated an age older than Tertiary, but beyond that Darwin did not venture a guess before his return to England. 'Perhaps some good conchologist', he wrote to Henslow, 'will be able to give a guess, to what grand division of the formations of Europe, these organic remains bear most resemblance.' This would give a lower limit to the age at which the Andes, or at least that part of them, had first been thrust up from the sea.

Darwin's *Gryphaea* is now termed *Aetostreon*, but falls within the same family. These oysters are characterized by two very unequal shells: the larger, curved, ridged and gnarly, has led to the common name Devil's toenail; the smaller, flattened shell formed a lid on top (see p.152). The animals lived in shallow seas, the large curved shell resting on the sea floor, the upper shell opening to allow the creature to filter-feed. D'Orbigny suggested to Darwin, by comparison with fossils from France, that the mountain-top shells were of Early Cretaceous age, and this is has been confirmed by recent research, placing them around 145–130 million years old.

After reaching the city of Mendoza, Darwin returned via a different pass, the Uspallata, where he chanced upon the fossil forest described in Chapter 3. In the space of three weeks Darwin had not only made the first geological survey of the central Andes, he had had the luck of a fossil-hunter's dreams. As he later wrote, 'Never did I more deeply enjoy an equal space of time.'

While Darwin was crossing the Andes, the *Beagle* paid a second visit to Concepción, and at Tomé in the north of the bay, the ship's assistant surgeon William Kent collected some fossils for Darwin. This time, Edward Forbes had first shot at the specimens, identifying three species (two bivalves and a gastropod) that d'Orbigny had previously discovered on the nearby island of Quiriquina. He also named two of his own: an ammonite and a nautiloid; the former is now known as *Eubaculites vagina*.

These are remarkable ammonites with a spear-shaped shell up to 6½ ft (2 m) long. They probably lived in the middle of the water column, feeding on small invertebrates. Forbes remarked on the similarity between the Tomé fossil and ones he had studied from India, of similar age, and Darwin concurred that 'this fact, considering the vast distance between Chile and India, is truly surprising'.

LEFT Fossils of the Late Cretaceous ammonite *Eubaculites vagina* from Tomé, Chile. The specimen includes two 1½ in (3.5 cm) fragments of the ribbed, tubular shell.

BELOW Reconstruction of the spear-shaped ammonite *Eubaculites*, which grew up to 6½ ft (2 m) long.

The *Nautilus* specimen comprised six segments of the curled shell. In a letter to Darwin, Forbes chivalrously suggested 'I propose to call it *N. D'Orbignyanus* if you have no objection'; this is still its specific name, now placed in the genus *Eutrephoceras* and corrected to *dorbignyanum* in accordance with current rules of nomenclature. D'Orbigny himself wrote to Darwin that since the bivalves and gastropod from Tomé were identical with the ones he had identified at Quiriquina, the deposits must be of the same formation; however, whereas he had taken them to be of Tertiary age, the addition of the ammonite and nautilus suggested an older, Cretaceous age. These deductions have been entirely borne out by recent research – the deposits at both localities are now known as the Quiriquina Formation, and are from the end of the Late Cretaceous, around 68–66 million years old.

Less than three weeks after his return from Mendoza, Darwin set off on his last major overland excursion in South America, starting at Valparaíso and proceeding via Coquimbo to Copiapó (see map p.19). The straight-line distance was only 420 miles (675 km), but Darwin and his guide spent two months exploring the region en route. On many days they made excursions into the lower valleys of the western flank of the Andes to study geology. The main fossil finds started around Coquimbo, from where on 21 May Darwin set out on a five-day excursion into the

The nautiloid shell from Tomé, *Eutrephoceras dorbignyanum*, named by Forbes in honour of Alcide d'Orbigny. At 2 in (5cm) in diameter, the individual was probably a juvenile; larger specimens measure up to 6 in (15 cm).

The living nautilus, a survivor of a once abundant group. Floating several hundred metres deep in the ocean, they are predators and scavengers of marine animals. The eye and tentacles are visible in this individual.

A cluster of rudist shells, showing their typical cone-shaped form and 'lid' on top. Rudists are bivalve molluscs but Darwin initially interpreted them as corals.

The central Andean region in the Late Cretaceous – volcanic islands with offshore faunas, remarkably similar to Darwin's vision of the Andes prior to uplift. Subduction of the Nazca tectonic plate (arrow) forces up the South American plate.

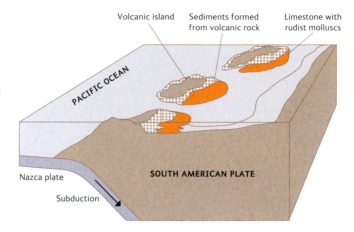

mountains. At the silver mine of Arqueros, a mass of limestone contained 'great numbers of Gryphaea and a large oyster', and fossils described by Darwin in his field notes as 'petrified tubiform corall' that in parts almost constituted the limestone. D'Orbigny later identified Darwin's coral specimens as *Hippurites chilensis*, a species he had named on the basis of very fragmentary material from the Arqueros region. D'Orbigny thought it was a brachiopod, but it is now recognized as a rudist bivalve.

Rudists were a very remarkable group of molluscs, characteristic of the Cretaceous. In the group of rudists to which Darwin's fossils belong, the main shell comprised a conical tube, with the second shell forming a lid on top. Rudists lived singly or in groups of individuals attached to the substrate, sometimes so densely that they were at one time considered equivalent to the reefs made by corals today. Given that these were among the first rudists to be described from South America and their true identity unknown at the time, it is unsurprising that with their 'tubiform' shape they were identified as corals by Darwin. The age of the Arqueros remains is now considered Middle Cretaceous, around 110 million years old.

Darwin then proceeded further up the valley to the River Claro (a tributary of the Elqui that flows to Coquimbo), as he had heard of a rich deposit of shells there. He was not to be disappointed: 50–60 ft (15–18 m) of rock were 'almost composed of an infinite number of a ribbed unequally valved shell', that he identified as *Terebratula* (a brachiopod). There were also pieces of *Gryphaea* and the outline of an ammonite. Forbes later identified the common brachiopod with *Terebratula aenigma* of d'Orbigny, now *Rhynchonella aenigma*; other, larger shells that he considered a new species are now identified as *Spiriferina rostrata*.

In places Darwin found the rocks solid with shells of the small brachiopod, *Rhynchonella aenigma*. These specimens, ½–1 in (1–2 cm) in diameter, are from the Jurassic of Huasco, Chile.

Brachiopods are marine invertebrates known as 'lamp shells', with double, hinged shells superficially resembling bivalve molluscs but only distantly related to them. They reached their maximum diversity and abundance during the Paleozoic, but persisted through the Mesozoic and are represented by around 400 species today. The identity of the brachiopods as an independent phylum of animals, separate from the molluscs, was still unclear at the time of the *Beagle* voyage.

Moving on toward Copiapó, Darwin found similar brachiopods at Guasco (now Huasco) and Vallenar, and at Amolanas where he had been directed to a rich shell-bed. Here it was the bivalve *Gryphaea* that was present in enormous quantities, in parts 'the rock being almost composed of them'. There were also brachiopods, a large gastropod and a fragment of ammonite. A lower bed contained 'thousands of a spiral univalve [gastropod].' This notebook entry not only shows Darwin describing a sequence of fossils from successive beds, but using them as a correlative tool in the field, concluding from the brachiopods and

A living brachiopod feeding with its shell valves open. Unlike bivalve molluscs, most brachiopods live anchored to the sea floor, extracting food particles from sea water using the coiled filtering device visible inside the shell.

Gryphaea in particular that the deposits here and at the Claro River were of the same age. Forbes later named two species from Amolanas after their finder: the oyster, *Gryphaea darwinii*, and a small, clam-like bivalve *Astarte darwinii*. Many of Darwin's fossils from this area are now considered Jurassic in age.

Staying at the house of one Don Benito Cruz at Copiapó, Darwin was amused to hear discussions concerning the nature of the fossil shells – 'whether they really were shells or had been thus born by Nature' – in other words, chance formations in the rocks. These were questions about the meaning of fossils which, Darwin noted, had been resolved in Europe a century before.

After resting a few days in Copiapó, and as the *Beagle* had not yet arrived, Darwin engaged another guide and a train of mules and set out in a northeasterly direction toward the mountains. In a valley named Despoblado ('uninhabited'), he found limestones with the familiar brachiopods and *Gryphaea*, and the following day climbed to around 10,000 ft (3,000 m) where he encountered 'much puna' and, in a deposit of sandstone, the same shells, again identifiable with those of the Claro River. Finally, on 12 July 1835, the *Beagle* sailed further north to the port of Iquique, then in Peru but now part of Chile. Darwin visited a silver mine at

Jurassic brachiopod shells, later appropriately named *Terebratula inca* by Forbes, collected by Darwin in Peru. The larger specimen is 2 in (5 cm) long.

Guantajaya, and in limestone deposits found a few fossil shells of a bivalve now identified as *Thyasira*. He also discovered a brachiopod later named *Terebratula inca* by Forbes and some probable fragments of ammonite. These were Darwin's last fossil finds in South America.

The uplift of the Andes

The fossil deposits of the Andes were intimately associated with volcanic rocks; uplift of the fossiliferous seabed and volcanic activity had, as Darwin supposed, gone hand in hand. In a diary entry while in the field, he commented, 'It is very interesting finding grand volcanic lava formation of age of ammonites.' As discussed above, there were later, Tertiary fossils within the Andes as well, so Darwin was in no doubt that the chains comprising the Andes had risen at different times. Our current understanding is that while different parts of the range have different histories, the central Cordillera that Darwin studied began to be elevated in the Late Cretaceous, some 100–70 million years ago. Much of its elevation, however, happened later, in the late Cenozoic, only about 20–8 million years ago.

In an important essay of 1840, Darwin invoked 'the grandeur of the one motive power, which, causing the elevation of the continent, has produced, as secondary effects, mountain-chains and volcanos'. In identifying a single underlying force Darwin was broadly correct, but whereas he saw that force as pressure of molten rock, we now know that it is the horizontal movement of massive areas of the Earth's crust, driven by the process of plate tectonics. In the case of South America, the Nazca Plate under the Pacific is expanding eastwards, colliding against the South American Plate (see p.148), and sliding beneath it. It is this process of subduction that has

at an elevation of 5000 feet & 45 leagues from the coast on the Andes Dept. of Copiapo.

The type specimen of *Gryphaea darwinii*, 3½ in (9 cm) long – one of the bivalve molluscs known as devil's toenails, with Darwin's original label.

pushed up the Andes and that leads to frequent earthquakes and, by melting of rock at the boundary, to volcanic eruptions. South America, in fact, is being squeezed from both sides, as the spreading of the Atlantic compresses it from the eastern side as well. The result is the general uplifting of the continent observed by Darwin along the entire southern coast, dramatically elevated along the Andean margin.

Yet periods of uplift had clearly alternated with those of subsidence, as Lyell had indicated and as Darwin had deduced from the thick, shallow-water, Tertiary deposits in Patagonia. The same must have happened during the formation of the earlier, Mesozoic deposits, since faunas that Forbes had told him lived in relatively shallow seas persisted through great depths of rock in the Andes.

The forces of subsidence and uplift must, Darwin believed, balance one another, so when the *Beagle* sailed from Lima into the Pacific, he pondered what evidence he might seek for the lowering of the ocean basin in compensation for the recent uplift of the South American continent. He found the answer in coral islands, as will be explained in the next chapter.

The oldest fossils in the world

Fossils older than Mesozoic, in rocks now considered Paleozoic in age, were relatively poorly-known in Darwin's day. The finds he made in the Falkland Islands were of great significance in this respect; as Darwin scholars Gordon Chancellor and John van Whye point out: 'It is difficult to overstate the importance of these fossils.'

On 1 March 1833, the *Beagle* sailed into the Falkland Islands, some 340 miles (550 km) east of the southern cone of South America, and stayed for around a month. Part of FitzRoy's instructions from the Admiralty had been to undertake a survey of the islands, especially their harbours; a second visit of similar duration took place almost exactly a year later. Darwin found the place dreary and 'with an air of extreme desolation', an impression not improved by the cold, driving wind and rain, but he pursued geology and natural history with his usual determination, and was to make two significant discoveries in the islands. The first was variation in the living dog-like species known as the Falkland fox, leading to his most explicit evolutionary statement of the entire voyage (see p.188). The second was the discovery of what were at that time among the oldest fossils discovered, equalling in age the most ancient specimens known from Europe.

After two weeks examining the geology of East Falkland, Darwin had found rather uniform slate rocks, with no trace of fossils. He had come to the firm conclusion that the rocks were so ancient that they 'belonged to that class which does not bear the signs of the coexistence of living beings with its formation'. They were, in the terminology of the time, Primary rocks, formed before the world became populated with living beings. Then, on 19 March 1833, at Port Louis in Berkeley Sound, he came across layers of more sandy slate, 'abounding with impressions and casts of shells'... 'The whole aspect of the Falkland Islands was... changed to my eyes from that walk'. Over three days, Darwin returned to the spot, collecting more specimens. In his *Geological Diary* he wrote, 'I think this is one of the oldest formations which even is fossiliferous', citing as reasons both the nature of the rock and the 'general character of the organic remains'. He makes a detailed comparison between the Falklands geology and that of ancient strata in Anglesey (North Wales) published by his mentor Henslow. Almost certainly, he was also drawing on his own experience of seeing ancient rocks and fossils of this kind during his excursion to North Wales with Adam Sedgwick in the summer of 1831, shortly before signing up for the *Beagle* voyage (see p.12). It is probably no coincidence that among his letters to Henslow, it is in one of the April 1833 that he writes, of Sedgwick: 'tell him I have never ceased being thankful for that short tour in Wales'.

The Falkland fossils themselves were mostly brachiopods; Darwin indicated his (correct) identification of them by naming them *Terebratula*, a living brachiopod genus with an extensive fossil record. He also realized that the Falkland shells must belong to several different species, writing that 'the shells all belong to Terebratula and its subgenera'.

A living stalked crinoid (sea lily) showing the stalk attached to the sea floor. The branched, feathery arms capture food and pass it to the mouth at the centre.

As well as brachiopods, Darwin correctly identified fragmentary stalks of sea-lilies (that he termed *Entrochites*), and mentioned that he had heard of one of its 'flower-like heads' having been found. Sea-lilies, like brachiopods, survive today as a shadow of their former abundance and diversity – around 80 species are known to exist, mostly inhabiting deep ocean trenches. They were at their peak in the Palaeozoic Era but were also common in the Mesozoic. Relatives of starfish, they attach to the substrate by a stalk, on top of which the main body bears 5–10 feathery arms used for trapping food particles. Feather stars are closely-related, more mobile forms that lack a stalk; together the two groups form the crinoids.

Many of the Falkland fossils, Darwin realized, were not the actual shells of the animals, or even mineral replacements of them, but natural impressions left in the rock as it solidified around them, the shell itself then dissolving away. The resulting moulds of the outer and inner surfaces were nonetheless faithful replicas of the originals, many of them stained a rusty colour that stands out against the dull brown rock. Darwin applied acid to the remains, confirming the absence of the original carbonate mineral, and wondered how it had escaped.

To Darwin and his contemporaries, these remains represented some of the oldest life on Earth, the rocks therefore belonging to the 'Transition' Period that was thought to follow the 'Primary'. Their significance was immediately clear to Darwin in the field: 'It will be pre-eminently interesting', he wrote in his diary, 'to compare these fossils with those of a similar epoch in Europe: to compare how far the species agree.'

Darwin sent several pieces of the precious fossil-bearing rock home in his next consignment, and collected more material on the *Beagle*'s return to East

Falkland a year later. Some specimens in the collection, however, are from West Falkland and were not collected by Darwin himself. In his *Geological Diary* in 1834, Darwin acknowledged 'a series of specimens which Mr Kent, when in the *Adventure*, had the kindness to collect for me at the Western island'. The *Adventure* was the vessel that FitzRoy had purchased to double-up the survey work; he placed it under the command of the *Beagle*'s First Lieutenant John Wickham, and Mr Kent was Assistant Surgeon. A further contribution came several years later, when Bartholomew Sulivan, who had been Lieutenant on the *Beagle*, visited the Falklands in 1844–1845 as Commander of HMS *Philomel*. At that time he collected several fossils for Darwin on Saunders Island, a small islet off West Falkland; the deposits there are of the same formation and age as those encountered by Darwin at St Louis.

Shortly after his return to England, Darwin showed the Falkland fossils to eminent geologist Roderick Murchison, who declared that both the rock and its included fossils were so similar to those of the Caradoc Sandstone of the Welsh

Falkland sandstone with brachiopods. One species, *Orthis* (now *Schellwienella*) *sulivani* (labelled c) was named after Bartholomew Sulivan; another (labelled b) is the fan-shaped *Spirifer* (now *Australospirifer*) *hawkinsi*. These Darwin specimens are known as the 'Royal brachiopods' since they were shown to Queen Victoria.

borders that the two could scarcely be distinguished. The Caradoc Sandstone was then placed early in the Silurian Period, a major interval of geological time that Murchison himself had identified and named only two years previously (today it is placed within the preceding Ordovician Period). Darwin seems also to have shown the fossils to James de Carle Sowerby (uncle of his mollusc specialist George Sowerby), who agreed with Murchison that the fossils 'decidedly belong to [the] old Silurian system', even considering some of the species identical to those found in Europe. Later, Murchison revised his age estimate somewhat, considering the Falkland finds Late Silurian or Devonian (the succeeding period). They are now accepted to be of Early Devonian age, around 400 million years old.

Formal study of the material was, however, assigned to John Morris and Daniel Sharpe, leading palaeontologists of the time, who identified no fewer than eight species of brachiopod among the samples, five of which are still considered valid. Ranging in size from around ¾ to 3½ in (2 to 9 cm), some such as *Schellwienella*, were finely ridged and roughly semicircular in shape when flattened on the rock; others, like *Australospirifer* were wider, fan-like, with more coarsely ribbed shapes.

As well as the new brachiopod species, Morris and Sharpe mentioned in passing that the rocks contained crinoid (sea lily) stems and 'fragments of a trilobite'. The latter was noted by Darwin in his *Journal of Researches* as 'an obscure impression of the lobes of a trilobite'. It was nonetheless the only fossil of this quintessentially Paleozoic arthropod group collected on the voyage. Within a slab containing predominantly brachiopods, and somewhat resembling them in its ridged structure, it is part of the tail of a trilobite of a group known as the calmoniids. These were bottom-dwelling creatures with well-developed eyes that probably preyed on or scavenged marine invertebrates. Four species of calmoniids have subsequently been identified in the Falkland rocks from which Darwin collected; his specimen can be identified from its shape as one of two species in the genus *Bainella*. These were large trilobites, around 5 in (13 cm) long when living.

An enigmatic aspect of Darwin's field description of his Falkland fossils concerns corals. In an entry in his *Geological Diary* that refers to his return visit in March 1834, he indicated that in one area of the deposit at Johnson Harbour, near to Port Louis, 'there were numerous casts which appeared to have been formed by some coral, such as *Gorgonia*'. In using this name Darwin was referring to branched, soft corals known as 'sea fans'. He indicated that they occurred 'in such quantity, that the rock is wholly composed of them', yet no specimens of coral can be found in Darwin's Falkland collection today, or

A Devonian scene, showing trilobites, *Walliserops*, related to the species found in Darwin's Falkland sandstone.

A tail impression of the trilobite *Bainella* in Devonian sandstone collected by Darwin in East Falkland. The tip of the tail is to the left of the box. Impressions of brachiopods can also be seen on the slab. The box width is 1 in (2.5 cm).

indeed in the fossil-rich deposits (now called the Fox Bay Formation) from which he collected. In 2015 the puzzle was solved when researchers compared Darwin's field specimen numbers with his labels on specimens in the Sedgwick Museum, Cambridge. The specimens in question are not corals but fragments of the branched arms of the crinoid.

As always, Darwin was thinking of the bigger picture when considering his finds. Geologists had suggested that fossils from different parts of the world resembled one another more closely the farther back in time we looked; regional differentiation was a modern phenomenon. In 1846, Darwin addressed this question with reference to his Falkland fossils, which by then had been shown to be not the same species as those collected in the Silurian and Devonian deposits of Europe. While acknowledging the general resemblance of ancient faunas across the globe, he pointed out that many of today's marine species are also very widespread, concluding that the idea of species having more global distributions in the distant past than at present 'must be greatly modified'.

There is a fascinating, apparently transmutationist (evolutionary) passage in Morris and Sharpe's account of Darwin's Falkland fossils. They wrote: 'The valuable researches of Mr Darwin have also revealed to us that the existing conditions of some portions of the southern hemisphere at the same era were favourable to the development of other species of the family Brachiopoda nearly [i.e. closely] related to those which in Northern Europe characterize the rocks of the Palaeozoic era.' Conditions favourable to the development of species closely related to others? A stronger hint at transmutation would be hard to imagine. That this was revealed to them by the 'valuable researches of Mr Darwin' suggests that the latter had felt able to discuss his transmutationist leanings with Morris and Sharpe, and that the idea had fallen on receptive ears.

Darwin also found the Falkland fossils suggestive of global changes in climate. While in the Falklands he suggested that the crinoids and other fossils might indicate a climate warmer than today. Later, he returned to the theme, pointing out that the English fossils similar to those of the Falklands were associated with others that he believed indicated a tropical climate, concluding that at that time most of the world had experienced warm conditions.

In a final twist, both the rocks and their fossil content in the Devonian of the Falklands were subsequently found to have more in common with those of South Africa than South America, with several of Darwin's species also found there. This circumstance is explained by continental drift, since the Falklands are now

Crinoid fossils from East Falkland. The ridged, tubular elements (top) are portions of stem, the thinner and more flattened pieces the arms, including branching fragments (e.g. at bottom centre) that led Darwin to identify them as corals.

known to have once been situated adjacent to the east coast of southern Africa, from which position they rotated 180 degrees into their current location.

When Africa and South America were conjoined in the Paleozoic Era they were also joined, via Antarctica, to India and Australia, forming the supercontinent Gondwana (see p.109). In Tasmania, Darwin found further fossils of the Gondwanan period, although they were later than those from the Falklands. Exploring the mountain foothills in the southeast of the island in February 1836, he encountered strata 'containing numerous small corals and some shells'. These included brachiopods as well as scallops and oysters (bivalves). Many were represented as casts; others were 'beautifully silicified'. Most of the specimens were collected by Darwin himself; others, from localities including the Huon River further inland, were given to him by the Surveyor-General, George Frankland.

The Tasmanian shells were passed to George Sowerby for study; six species of brachiopod were identified, all of them new, in two genera (*Producta* and *Spirifer*). In addition, there were various 'corals'; these were described by William Lonsdale, a specialist of coralliform fossils, as six new species in three genera. All of the Tasmanian 'corals' are now identified as bryozoans; the two groups were only understood to be fundamentally distinct in the 1830s. Nonetheless, the specialists recognized that all the fossils were similar to those found in 'Silurian, Devonian and Carboniferous strata of Europe… all having undoubtedly a Palaeozoic character'.

The Tasmanian deposits from which Darwin collected are now placed within the Permian, the last period of the Palaeozoic Era, spanning 299–252 million years ago. For example, on Maria Island, an islet off the west coast of Tasmania, Darwin had found 'limestone almost composed of parts of bivalves'; this deposit is now known as Darlington Limestone and is well-known for its richness

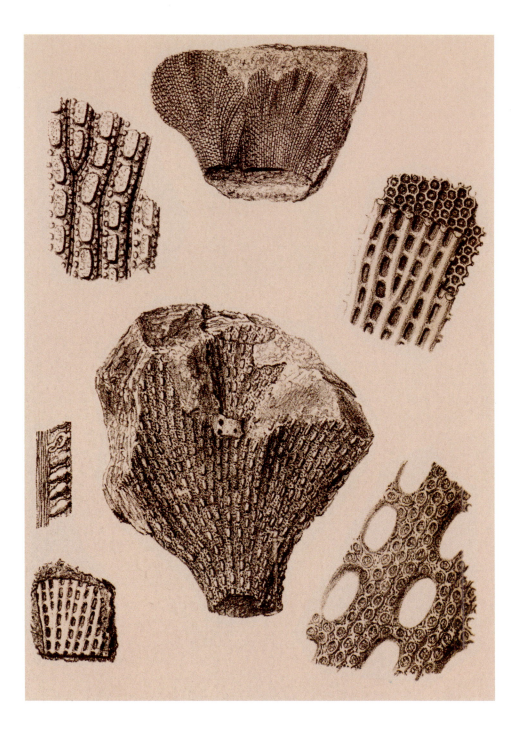

ABOVE Brachiopod shell moulds collected by Darwin from the Permian of Tasmania. Left, 1 in (3 cm) shell of *Terrakea* and right, the massive 5½ in (14 cm) fan-shaped shell of *Spirifer vespertilio* with fragments of *Ingelarella* and *Licharewia* species alongside.

OPPOSITE Darwin's bryozoan fossils from the Permian of Tasmania. Top and top left, *Fenestella fossula* (and close-up); top right, close-up of *Hemitrypa sexangula;* bottom, *Parapolypora ampla* (and three close-ups).

in the Gondwanan scallop *Eurydesma.* The Permian had only been named in 1841, based on rock exposures in Russia, so the broad conclusion of Darwin's collaborators that the Tasmanian fossils were Paleozoic was reasonable enough.

Darwin's bryozoan fossils are now lost, but from the illustrations and Lonsdale's descriptions, several of the species were fenestellids, bryozoans forming fan-shaped colonies that stood erect in the water attached to the substrate by a holdfast. Other species, stenoporids, formed more trunk-like, branching colonies.

The brachiopods are also all Permian species, many of them known only or mainly from Australia. Within the two preserved blocks in the Darwin collection, no fewer than six genera have been recognized. The blocks also contain a gastropod mollusc and fragments of crinoids (sea-lilies) and ostracods (tiny shelled crustaceans).

John Morris, who with Daniel Sharpe had named Darwin's Falkland brachiopods, also examined his Tasmanian collection, as an aid to identifying fossils collected by explorer and geologist Paul Edmund de Strzelecki. From among Strzelecki's collection Morris named a new species, *Spirifer darwinii,* commenting: 'I dedicate this species to Mr. C. Darwin, who has very largely contributed to the advancement of physical geology and natural history generally.'

CHAPTER 5

∼

Coral islands

THE NATURE AND FORMATION of coral reefs were the subject of much debate in the early 19th century, but it was a problem to which Darwin would make a major contribution during the course of the *Beagle* voyage. Reefs are built mainly by colonies of numerous tiny animals, related to jellyfish and sea-anemones, which produce a hard, limestone exoskeleton to support and protect themselves. Those that rest on the shallow sea floor close to land are known as fringing reefs (a term introduced by Darwin himself). Barrier reefs are more massive structures, resting at greater depths further from the shore. Both of these can be found along continental land masses and also around islands. The third category, the atoll, is a ring-shaped coral island or chain of islands emerging from the deep mid-ocean.

OPPOSITE Polyps of the reef-building coral *Galaxea*. Darwin marvelled at the 'mountains of stone accumulated by the agency of various minute and tender animals'. Each polyp is some ¼ in (5 mm) in diameter.

RIGHT A typical coral atoll, a loop of land in mid-ocean, with white breakers crashing against the outer reef and a shallow, relatively calm lagoon inside. The origin of these structures was a mystery.

The origin of atolls in particular had been a source of wonder. French naturalists, Joseph Gaimard and Jean Quoy, had shown in 1825 that reef-building organisms live only in relatively shallow water, so that by the time of Darwin's voyage it was generally accepted that atolls must rest on a submarine mountain or volcano. In the second volume of his *Principles of Geology*, published in 1832, Charles Lyell promoted the theory, first suggested by British navigator William Beechey, that the circular or ovoid shape of many atolls resulted from their resting along the rim of an extinct, submerged volcano. The shallow inner lagoon typical of an atoll overlay the volcanic crater.

For all his admiration for Lyell's work, Darwin considered the crater theory of atolls a 'monstrous hypothesis', for several reasons. First, some atolls are too large to reflect a volcanic crater – Darwin cited known examples up to 60 miles (100 km) across. Second, in many cases atolls have a very irregular shape, far from the circle or oval expected from a typical crater. Third, and more subtly, Darwin realized that it would be an extraordinary coincidence if so many submarine volcanoes across the oceans had stopped growing at roughly the same height so as to place the reef-building organisms within their narrow, shallow zone of tolerance.

A different hypothesis was needed, and it grew from Darwin's observations on the west coast of South America (see Chapter 4). Darwin was convinced that the massive uplift of the continents, which he had seen so strikingly exemplified in the Andes, must be compensated by subsidence elsewhere, almost certainly in the oceans. In this Darwin was following Lyell, but it led him to a radical explanation for the origin of coral atolls. Many volcanic islands were known in the world's oceans; Darwin had studied several of them himself during the course of the voyage. An emergent volcano, forming an oceanic island and then becoming extinct, would support the growth of a fringing reef around its shores, as seen in many instances. But if the ocean floor then began to subside, the volcano would sink down and the corals attached to its flanks would find themselves in progressively deeper water. Provided the subsidence was slow enough, however, the coral organisms, by building their limestone foundations upwards, could keep up with the deepening water and remain in their preferred shallow-water zone. At the same time, as the emergent volcanic peak became progressively smaller, the reef, growing vertically upward from its foundation, found itself further from the shore – the fringing reef had become a barrier reef. Finally, as the peak of the volcano disappeared beneath the waves, and the coral continued to grow upwards, it was left as a loop-like island whose shape reflected the former coastline (as opposed to the crater rim) of the sunken volcano – an atoll.

Darwin's theory for the origin of an atoll. Left, a fringing reef (A) grows on the flanks of an oceanic volcano; centre, as the volcano sinks, the corals grow upward to form a barrier reef (A'); right, the volcano has sunk beneath the waves and further coral growth has formed an atoll (A").

Late in life, Darwin wrote that he had conceived the essence of his coral theory while still in South America. It was, he said, thought out in a 'deductive spirit', for he had yet to even see a coral reef. His notebooks written in Chile in 1835 contain only fragmentary, somewhat ambiguous comments on the subject, which has led some scholars to question Darwin's memory on this point. There is no doubt, however, that as the *Beagle* set sail across the Pacific he was already thinking deeply about the question, and was looking forward to his passage across the Pacific and Indian Oceans where the vast majority of the world's atolls occur.

Across the Pacific

First port of call, in September 1835, was the Galápagos Islands, where Darwin found much else to occupy him, including a few fossil shells embedded in volcanic deposits, but he saw no coral reefs. It was known that reefs were largely tropical in distribution; few occurred beyond latitude 30 degrees north and south of the equator. The Galápagos, however, straddled the equator – why were there no reefs? Darwin acknowledged Captain FitzRoy for suggesting the answer – the water around the islands was too cold. Subsequent research has confirmed that most reef-building coral animals require water temperatures of at least 60°F (15°C), and are therefore largely excluded here because of the cold water current that runs up the west coast of South America and bathes the Galápagos. Only on the northernmost island of the group, not visited by the *Beagle* but where ocean temperatures are a few degrees warmer than in the main archipelago, do coral reefs flourish. It is serendipitous that the island in question was later named Darwin Island.

As the *Beagle* made her way across the Pacific, Darwin had his first view of a coral atoll, then termed a lagoon island. In November 1835 the ship passed through the Low Archipelago, now the Tuamotu Islands and part of French Polynesia. Climbing the masthead, Darwin saw a 'long and brilliantly white beach' and a 'wide expanse of smooth water within the ring', startling in its contrast to the ocean waves all around.

A few days later the *Beagle* arrived at Tahiti and remained for 12 days. Here Darwin witnessed an island that was mountainous but encircled by a barrier reef around much of its perimeter. Between the reef and the shore was 'an expanse of smooth water, like that of a lake'. The decisive moment came when Darwin ascended one of the peaks and gained a clear view of the neighbouring island of Moorea (then named Eimeo). As on Tahiti, the mountains of Moorea rose out of a calm lagoon, encircled by a ribbon of coral reef. Darwin was struck with the realization that there was no essential difference between the reef and lagoon of an atoll, and those of a volcanic island with a fringing or barrier reef.

The island of Moorea surrounded by a barrier reef. Darwin viewed the island from Tahiti and was forcibly struck by its similarity to an atoll.

'Remove the central group of mountains', he wrote, 'and there remains a Lagoon Island'. Darwin had not only explained the origin of atolls, but he had done so by showing that fringing reefs, barrier reefs and atolls were stages in a continuous sequence of change.

While on Tahiti, Darwin hired a canoe to observe the delicate corals growing in the lagoon. He did not reach the outer edge of the reef, but local people told of the massive corals growing on the ocean side, and how storms would tear off chunks of reef and throw them inland. They further explained these were quite different species from those growing in the quiet of the lagoon. Darwin also studied as many maps and published accounts of coral islands as he could find in the *Beagle* library. He now had enough information to flesh out his theory, and as the *Beagle* sailed from Tahiti to New Zealand he wrote it out as a 22-page essay, his first extended treatment of a theoretical subject.

Having speculated that the barrier reefs of Tahiti, and the atolls of Tuamotu, might show stages in a sequence of development, Darwin witnessed an apparently intermediate stage as the ship passed the island of Whytootake (now part of the Cook Islands) on 3 December 1835. The central hilly island was surrounded by a circle of reefs (as on Tahiti), but the reefs had been 'converted into low narrow strips of land' consisting of 'sand and coral rocks heaped upon the dead part of a former reef' (as in an atoll). It was, he remarked, a 'union of the two prevailing types of structure'.

The Cocos (Keeling) Islands

The opportunity to put his ideas to the test came after the *Beagle* left Australia in March 1836 to cross the Indian Ocean, and stopped for 12 days at what are now known as the Cocos (Keeling) Islands. The intensive work undertaken by the young naturalist and his nautical hosts at Keeling Atoll, the main island of the group, was the clearest confluence of their interests in the entire voyage. FitzRoy, as much as Darwin, was well aware of the scientific interest surrounding atolls. Moreover, there was a strong practical purpose in their study, as on the one hand their lagoons provided respite for ships in mid-ocean; but on the other they were the cause of many nautical accidents as ships became beached or wrecked on coral rocks that were barely visible above the waves or submerged just below them. FitzRoy's instructions from the Admiralty at the start of the voyage had included a request to study the detailed structure of one or more atolls. They also suggested a possible stop at the Keeling Islands to accurately determine their position, although this assumed that

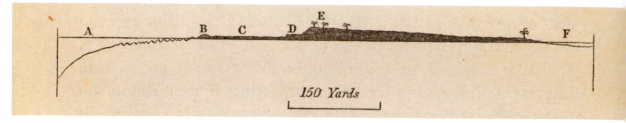

Darwin's idealised cross-section of one side of Cocos (Keeling) atoll, based on a series of transects. The ocean side is at A, the inner lagoon at F.

the *Beagle* would pass north of Australia into the Indian Ocean. In the event she sailed south of Australia and Keeling was then a significant detour to the north; it has been suggested that Darwin himself may have influenced FitzRoy's decision to call there nonetheless.

The main island of the Keeling group, which the *Beagle* reached on 1 April 1836, is a horseshoe-shaped atoll, some 10 miles (16 km) in length, comprising 24 islets around a shallow lagoon. With the *Beagle* at anchor in the lagoon, and many of her crew engaged on surveying the atoll, Darwin set about investigating its structure. He made detailed personal observations on at least two of the islets, and gleaned valuable information on others from the officers engaged in the survey, especially Bartholomew Sulivan. Darwin's method was to construct what became known as a transect – a series of observations at intervals along a straight line, in this case from the ocean to the lagoon – and his work on the Cocos (Keeling) Islands is considered pioneering in this regard. In the process he collected a series of specimens that illustrated the transition from living coral to the calcareous rock that formed the reef, and were instrumental to his developing ideas about atoll growth and structure.

Determined to examine the outer edge of the reef where it met the open ocean, Darwin 'reached far into the breakers... by the aid of a leaping pole' (between A and B on the transect above). In this, as in much of his other field work, he was very likely accompanied by his servant Syms Covington. Here they saw several species of living coral, in three principal growth forms (see facing page). Small pieces were hacked from each type and brought home. Darwin noted two important features of the growing edge of the reef. First, the corals were very sensitive to exposure to air – on some of the larger masses, the upper

BELOW Darwin's 3½ in (9 cm) sample of one of the large, rounded masses of living *Porites* coral from Keeling atoll, up to 8 ft (2.5 m) across.

RIGHT Close-up of the calcareous skeleton of the *Porites* coral. Each 1 mm depression (known as a corallite) housed an individual living polyp.

RIGHT Darwin's 8 in (20 cm) sample of the branch-like coral *Acropora* from Keeling atoll. He labelled it *Madrepora* – the name then in use for this and related genera.

BELOW Behind the outer reef, Darwin found a ridge formed from the calcareous skeletons of coralline algae, and collected this 10 in (4 cm) sample.

ABOVE A 5 in (12 cm) sample of living coral from Keeling atoll, this time formed of thick, intersecting plates and correctly identified by Darwin as a species of *Millepora*.

surface had died through occasional exposure at low tide, while the sides were alive. Second, the growth forms of the corals in several places convinced him that the reef was extending seaward, having reaching the limit of its possible growth in an upward direction.

Encrusting the dead parts of the coral were what Darwin called Nullipora or Corallina, now termed coralline algae. These are primitive plants, belonging to the species of red algae that are hardened by calcareous deposits in their cell walls, and are an important contributor to the growth of coral reefs. Darwin noted their pink colour, and that they were more tolerant of exposure than the corals themselves, forming banks up to 3 ft (1 m) high behind the reef edge and covered only at high tide (B on the transect). Underneath the living algae, Darwin found the reef excessively hard, requiring exertion with pickaxe and chisel before he 'at last attained a fragment & strongly suspect it is Corallina petrified'.

Further back, behind the algal ridge, no coral or coralline algae grew, except in occasional holes or channels. Instead, a remarkably flat, solid floor of coral rock was exposed, some 100-300 yards (90–275 m) wide, where again 'I could with difficulty, and only by the aid of a chisel, procure chips of rock from its surface' (C on the transect). Proceeding further inland the surface was raised a few feet (D), and was clearly formed from the accumulation of bits and pieces of coral cemented together, along with hard parts of other animals such as the shells of molluscs and spines of

Fragments of dead *Acropora* coral collected by Darwin inland on Keeling atoll and showing stages of erosion. Left, 5 in (12 cm) specimen with original columns partly intact; below, 4 in (10 cm) specimen after further erosion, only the bases of the columns remain.

Specimens of reef rock (that he termed 'coral rock') collected by Darwin on Keeling atoll. Left, 3½ (9 cm) example showing bits and pieces of corals, mollusc shells and the hard parts of other marine organisms cemented together; right, 3 in (8 cm) example where the original organic structures have been obliterated by dissolution and recrystallization of the skeletal minerals into a solid mass. The darker upper surface may be the original surface of the reef flat.

sea urchins. Further inland still, the highest areas of the island (E) were formed from lumps of coral, some of them rounded by the action of the tides, cemented together to greater or lesser degrees, as well as quantities of coral sand.

Darwin collected a series of specimens, ranging from relatively unaltered, identifiable pieces of coral, to branches that had been worn down by the action of the tides. Some specimens retained their original fine structure, others were petrified to varying degrees, including some 'where it was impossible to discover with the naked eye any trace of organic structure'. They showed stages, in other words, in the process of becoming fossilized.

Loose pieces of coral, 1¼–2¾ in (3–7 cm) long and worn to pebbles by the action of the sea, collected by Darwin on Keeling atoll. He anticipated that such pieces would become cemented together to form conglomerate.

For Darwin the origin of all this material was clear. Pieces of fresh coral could be seen strewn along the reef flat and sometimes further inland. Their origin had to be from the outer reef, broken by the action of the sea and 'thrown up by gales of wind & heavy surf & spring tides'. Mr Liesk, an English resident of the island, told him that shells contained within the fragments were mollusc species that lived only on the ocean side. The accumulated debris had subsequently become, as Darwin observed, cemented together by calcium carbonate precipitated from percolating water. Even on the reef flat, Darwin noted pieces of recent coral wedged in crevices, and envisaged their eventual incorporation by cementation and smoothing over by the tide. He also suggested that finer coral sediment, forming beaches and dunes on the island, might be the product of fish (such as parrot-fish) that scrape coralline algae with their hard beaks and excrete it as sand. Darwin credited this idea to Captain FitzRoy, and it has subsequently been confirmed.

The final stage of the transect was the lagoon (F), where corals of a quite different kind grew. Darwin recorded at least half a dozen different living species; they were 'elegant and more openly branching structures', but all were brittle and soft, much less heavily calcified than those of the outer reef.

Seaward beach on Cocos (Keeling) atoll in the 1970s, strewn with pieces of coral thrown up from the outer reef. Darwin recognized these were the source of the reef rock and conglomerate that built the island.

Miniature sheet of postage stamps issued by Cocos (Keeling) Islands in 1981, showing the 'saucer-shaped' form of the atoll. Above is a passage written by Darwin in 1836, beginning 'I am glad we had visited these islands...'

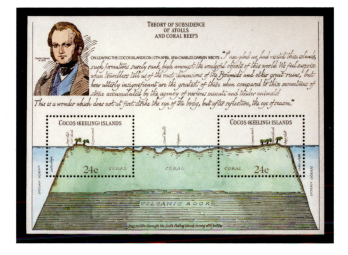

His observations, as a whole, illustrated the saucer-like shape of the atoll, but his explanation for this did not depend on its cresting the rim of a volcanic crater. He had confirmed the suggestion of German naturalist Johann Eschscholtz that reef-building corals and coralline algae grew most vigorously where they were exposed to the nutrient- and oxygen-rich ocean waters, forming the outer edge of the saucer. Moreover, Darwin conjectured that each time the land subsided a little, 'the outer parts near the Beach would constantly be repaired, but not so the interior'. Finally, once they reached the surface, the outer corals had a tendency to extend seaward, leaving behind them only coral rock, the living animals having perished due to periodic exposure to the air. This was the origin of the extensive reef flat behind the growing edge. Equally important, it was clear that the delicate corals that grew in the still, sediment-rich waters of the lagoon were not those that built the solid reef.

Darwin also felt he had the answer to a potential objection to his subsidence theory – the projection of many atolls above the level of the sea. The higher ground had not grown in place but was built of material thrown up from the living reef. He recognized that this must have taken time, so he suggested that subsistence was episodic and that an atoll such as Keeling must have been for some time in a resting phase. The extent to which the reef had grown outwards, and the width of the dead reef flat left behind it were, he concluded, a measure of the time since the last episode of subsidence. He even allowed that emergent atolls might reflect intervals of minor uplift within the overall trend of subsidence.

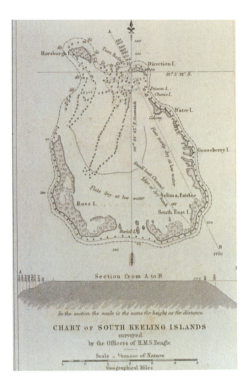

Keeling atoll, based on the survey completed by the *Beagle* in 1836. The atoll consists of a ring of islets around a central lagoon. The *Beagle*'s deepest sounding of 1,200 fathoms (7,200 ft or 2,200 m) is seen at B.

All of these observations were consistent with the subsidence theory, but they did not prove it. However, Darwin believed he had found 'tolerably conclusive' evidence of recent subsidence on the island itself. Around the lagoon, coconut trees that must have originally grown on dry land were being undermined by the lapping waters. And FitzRoy pointed out 'the foundation-posts of a storehouse which the inhabitants said had stood, seven years before, just above high-water mark, but now was daily washed by the tide'. Nowadays these observations would be explained by natural variations in sea level due to factors such as wind, ocean temperature, or a storm surge caused by a low-pressure weather system. The overall height of the Cocos (Keeling) Islands has not changed appreciably since Darwin's time; as will be discussed below, subsidence takes place on timescales of thousands to millions of years.

Much more significant were the soundings taken by the *Beagle*'s crew off the coast of Keeling Island. Darwin took intense interest in these, jotting down figures and rough calculations in his field notes. They were effectively an extension of his transect down the side of the reef, invisible beneath the waves. Soundings were commonly used to measure the depth of the ocean beneath a ship by means of a lead and line. Darwin and the *Beagle*'s crew also used the soundings to investigate items on the seabed, using a bell-shaped lead about 4 in (10 cm) wide with a concave underside that was plastered with a preparation of tallow. Loose items adhered to this, or if it hit something hard an impression would be left. With each sounding, Darwin wrote, the tallow was 'cut off and brought on board for me to examine'. He recorded some 40 such attempts, variously recognizing on the tallow fragments of, or impressions of, sand, dead coral, living coral, coralline algae, molluscs, sponges and other marine organisms. A clear pattern emerged as they moved further below the surface. Down to 48 ft (15 m) and probably to 72 ft (22 m) the tallow was marked with living coral; between 72 and 120 ft (37 m) there was sand and indentations of dead coral; and beyond that only sand. Darwin had confirmed and refined the key suggestion by Quoy and Gaimard, some 12 years previously, that reef-building corals live only in relatively shallow waters.

Equally important was confirmation of the great depth of the ocean around the atoll. In an exceptionally long sounding for the time, FitzRoy found no bottom with a line of 7,200 ft (2,200 m), just over a mile (1,600 m) offshore. By simple trigonometry the side of the island had to be at least as steep as 48 degrees which, citing Humboldt, Darwin noted was steeper than any known volcanic cone. The atoll could therefore not be merely a shallow cap to a volcano reaching close to the surface. Further evidence came when some of the lines let down close to the reef were cut, 'as if rubbed', at 3,000 and 3,600 ft (900 and 1,100 m) 'indicating the probable existence of submarine cliffs' with sharp edges of dead coral.

The evidence, although circumstantial, pointed to a great column of coral rock that had accumulated over a long period, and since the corals only lived and grew in shallow water, the implication was that they had grown upward to keep pace with a slowly sinking volcanic base. Darwin recognized, however, that as no individual coral colony grows to more than a few feet in height, the deep column must reflect 'the successive growth and death of many individuals', each 'being broken off or killed by some accident' and another growing on top of it.

The global picture

Leaving Cocos (Keeling) on 12 May 1836, the *Beagle* crossed the Indian Ocean to the island of Mauritius, where Darwin fortuitously completed his set. Having examined a barrier reef at Tahiti and an atoll at Keeling, here was a fringing reef close to the shore of a volcanic island. The officers had no need for a survey, so

Darwin took matters into his own hands and, with an unidentified accomplice, took a boat and the survey equipment to the seaward side of the reef, about half-a-mile off the western side of the island. 'At each cast', he wrote, 'we pounded the bottom with the lead' (lined with tallow). The results mirrored those at Cocos (Keeling) – down to 90 ft (27 m) an array of reef-building coral species was recorded. From 90–120 ft (27–37 m) there were quantities of the delicately branching coral *Seriatopora* - not, he noted, a kind 'efficient in forming a reef'. Beyond that the bottom was mostly sandy, and at the deepest sounding, at 500 ft (160 m), there were only 'bits of dead coral and a volcanic pebble'. He also recorded, as at Keeling, pieces of coral forming, and thrown on to, the island's beaches. The fundamental similarity to an atoll, in so different a geographical situation, confirmed Darwin's hunch that these were stages in the same process.

In the company of Captain John Lloyd, the Surveyor-General of Mauritius, Darwin explored the interior of the island, on one occasion from the back of Lloyd's domestic elephant. In many places he saw exposures of coral rock, to a height of 30–40 ft (10–12 m) above sea level, well above the reach of present-day tides. The beds were 'composed of precisely the same materials such as now on the beach… partially cemented fragments of strong branching corals'. In another place there was an extensive flat of hard, compact coral rock, and rising from it, two hillocks of coral rock 20 ft (6 m) high, their upper part containing large, recognizable blocks of coral that Darwin identified as *Astraea* and *Madrepora*. He collected a series of representative specimens of which only one (a piece of limestone) appears to survive. The conclusion was inescapable: the island had risen in relatively recent times; the various uplifted coral deposits could be identified as ancient coral flats and conglomerates that had formed closer to sea level; and the hillocks, Darwin felt, were probably areas of formerly living reef uplifted intact. Moreover, although there were no historical records of eruptions on Mauritius, cones and lava flows spoke of relatively recent volcanic activity.

Through all his observations on coral reefs, the global picture was never far from Darwin's mind. On Keeling Island, the inhabitants told him they had experienced three earthquakes in the previous 10 years. This, for Darwin, was linked to the island's subsidence, but it spoke of something of wider extent. Already, on a back page of his notes, he had scribbled 'Cocos Isd. connected with volcanic force of Sumatra. That rises, this falls'. This was a direct analogy with his observations in South America – the rise of land with the earthquake he had witnessed at Concepción would be compensated by a corresponding drop in the ocean floor of the adjacent Pacific Ocean. Similarly, an earthquake associated with uplift of the island of Sumatra, at the eastern edge of the Indian Ocean, would, Darwin believed, be felt at the same time on Keeling Island in the adjacent mid-ocean, but with a corresponding downward movement.

On Mauritius, uplift was evidently associated with volcanic activity, and in an exhaustive compilation of known facts about coral reefs all round the world, completed after his return to England, Darwin showed that the pattern was far from random. Islands with fringing reefs were often volcanically active or recently so, and were either stable or showed signs of recent uplift. Atolls, and islands with barrier reefs, were built on extinct volcanoes, and when plotted on a map of the world tended to be clumped in mid-oceanic areas of presumed subsidence. It was a brilliant piece of evidence in support of his theory, the two categories representing earlier and later stages in the process.

Darwin's theory put to the test

On his return to England, Darwin told Lyell of his new theory. Lyell 'was so overcome with delight that he danced about and threw himself into the wildest contortions, as was his manner when excessively pleased'. Instantly abandoning his crater-rim theory, Lyell wrote to Darwin, 'I could think of nothing for days after your lesson on coral reefs, but of the tops of submerged continents. It is all true, but do not flatter yourself that you will be believed, till you are growing bald, like me.' He arranged for Darwin to present his findings at a meeting of the Geological Society of London on 31 May 1837, at which Darwin very likely used some of his specimens from Cocos (Keeling) and Mauritius as visual aids. The theory became more widely known when Darwin included an extended

Sir Charles Lyell, whose geological insights inspired Darwin on the *Beagle* voyage. He 'danced about' on hearing of Darwin's theory for the origin of coral reefs.

summary of it in his *Journal of Researches* in 1839; and in 1842 it became the subject of his first scientific book, *The Structure and Distribution of Coral Reefs*. Acceptance of the theory was aided by Lyell's adoption of it in the next and subsequent editions of his *Principles of Geology*. The book that had been Darwin's 'bible' throughout the *Beagle* voyage was now being revised in the light of his own discoveries.

Not all geologists accepted Darwin's theory, however, and in the later decades of the 19th century many turned away from the idea of subsidence, suggesting instead that atolls grew on platforms either built up by accumulation of sediment on deep volcanic foundations, or eroded down from emergent volcanoes by the action of the sea. A particular version of the latter idea was prompted by the theory of ice ages. In 1837, just as Darwin was announcing his coral reef theory to the world, Swiss-American geologist Louis Agassiz proposed his glacial hypothesis, suggesting that vast sheets of ice had once covered much of the northern hemisphere. A corollary of the ice age was a substantial drop in the global sea level, since so much of the world's water was locked up in the expanded ice sheets. The potential impact on coral reefs was explored particularly by Canadian geologist Reginald Daly, who in 1915 published his glacial control theory of atoll formation. Daly proposed that during the glacial period, with the sea level some 160–300 ft (50–90 m) below present, the exposed tops of oceanic volcanoes were eroded away by the action of the waves, leaving a platform at around sea level. When the glaciers melted and sea-levels rose, coral reefs grew upward from these platforms to form present-day atolls. The latter were therefore all of relatively recent age and not more than around 160 ft (90 m) deep; moreover, there was no need to invoke Darwinian subsidence.

Darwin and Lyell had been well aware that the subsidence theory would be proved only by the discovery of great depths of coral strata resting on volcanic rock, either below an existing atoll or in ancient marine sequences exposed on land. Darwin had proposed as early as his 1835 essay that such a discovery could be explicable only by subsidence and a compensatory upward growth of the reef, given that reef-building corals live only in shallow water. As he later wrote to Alexander Agassiz (geologist son of Louis), 'I wish that some doubly rich millionaire would take it into his head to have borings made in some of the Pacific and Indian Atolls.' Alexander was a proponent of the sedimentation and erosion theory, and such a boring would decide between their respective theories. He was also, as Darwin knew, more than doubly rich, but he didn't take the hint and it was left to the Royal Society, in the late 1890s, to attempt the task.

The top of the borehole on Enewetak atoll that in 1952 finally demonstrated Darwin's theory. In 1976, Brian Rosen led an expedition that rediscovered the borehole and erected a suitable plaque.

They chose Funafuti atoll (in modern Tuvalu) and drilled 1,114 ft (340 m) down through limestone deposits. But it was not certain that the deeper limestone had been formed by reef-building corals, and the core failed to reach volcanic bedrock, so the result was inconclusive. Only in 1952, when the US government drilled deep boreholes in the Pacific atolls of Bikini and Enewetak in preparation for atomic bomb testing, was volcanic basement reached below some 4,000 ft (1,200 m) of shallow-water coral rock. The coral rock had, moreover, been forming for over 50 million years. Darwin's theory had been corroborated, 117 years after he first proposed it, and subsequent drilling through other atolls, as well as seismic profiles, have shown that subsidence was widespread.

Nonetheless, ice-age sea-level changes did have a major impact on coral reefs, as they did on all coastal environments. This geologically recent factor is now understood to be superimposed on the much longer-term process of subsidence, rather than an alternative to it. Moreover, whereas Daly envisaged a single glacial period several hundred thousand years in duration, we now recognize multiple glacial-interglacial cycles, each associated with major oscillations in sea level. Coral reefs were alternately submerged below sea level and exposed far above it, so had to re-form with each cycle.

In the last glaciation, the global sea level fell by some 400 ft (120 m), exposing former reefs high above the water mark and doubtless eroding their base of coral rock to some extent. Reefs would have started to grow on lower platforms, provided the water was warm enough. Then, following the last glacial maximum some 21,000 years ago, the sea rose steadily, and when it flowed over the elevated former reef platforms, corals began to grow on them again. At the same time, reefs that had formed at a lower sea level now began to grow upwards, following the rising sea level, but if the rise was too rapid they drowned. The point at which corals can grow just fast enough to keep up with a rising sea level has been termed the Darwin Point.

On Keeling atoll, radiocarbon dating of fossil corals in the conglomerate shows that they were growing between 4,000 and 3,000 years ago. Sea level then fell away slightly and the surface corals were exposed and died. The fossil coral and solidified conglomerates studied and collected by Darwin were all formed around this time, and are now seen as the remains of a former reef-flat. Above it, sandy deposits and looser coral rubble have built up more recently by the action of wind and waves, just as Darwin envisaged.

Deeper under Keeling atoll, fossil reefs some 120,000 years old have been discovered 20-45 ft (6-14 m) below present sea level. This dates them to the interglacial (warm period), before the last glaciation, when sea levels were similar to, or a few metres higher than, those of today. The time period since these corals formed is long enough for us to observe Darwinian subsidence, while allowing for uncertainties in the height of past sea levels and ancient erosion of the reef surface. An estimated last interglacial sea level 20 ft (6 m) above present, added to the 20 ft (6 m) depth of the fossil reef, gives roughly 40 ft (12 m) of subsidence in 120,000 years. This represents just under half an inch (1 cm) per century, so it is not surprising that the lowering of the island since Darwin's day is barely noticeable. The fossil reefs seen by Darwin at 30-40 ft (10–12 m) elevation on Mauritius are also probably from the last Interglacial period, so given the high sea level of that time, may not have been uplifted quite as much as he imagined.

A final piece of the jigsaw has come with the theory of continental drift, and its driving force, plate tectonics, which explains why areas of the ocean floor are subsiding. Magma from beneath the Earth's crust wells up to form mid-ocean ridges, on either side of which the newly formed ocean floor slowly spreads outwards. Volcanoes often form in this highly-active area, but as the ocean floor cools and contracts, it sinks down, taking volcanic islands with it. The islands are thus not only sinking but moving horizontally as the sea-floor continues to spread. Moreover, where the leading edge meets another continental plate, it slides beneath it and mountains may be thrown up in the process. Darwin's hunch, that the atoll question was linked to processes at a global scale, was correct in principle. He had imagined subsidence and elevation of adjacent areas as a kind of lever process, but would have rejoiced at the discovery of its true cause. As for the Cocos (Keeling) Islands, volcanic rocks have recently been dredged from nearly 2 miles (3 km) below the ocean surface, and have been dated to some 50 million years ago. The islands must have formed around this time, as part of the sea-floor activity that was separating India from Australia. They then subsided as the ocean floor moved northward, taking the islands with it, while the leading edges of the plates sank beneath and uplifted parts of Southeast Asia, including Sumatra.

Map showing the position of the Cocos (Keeling) Islands (green star) at the time of their formation some 50 million years ago. The red line and the I-A transform line mark the boundary between the Indian and Australian plates. Compare position of Cocos (Keeling) with map on p. 13.

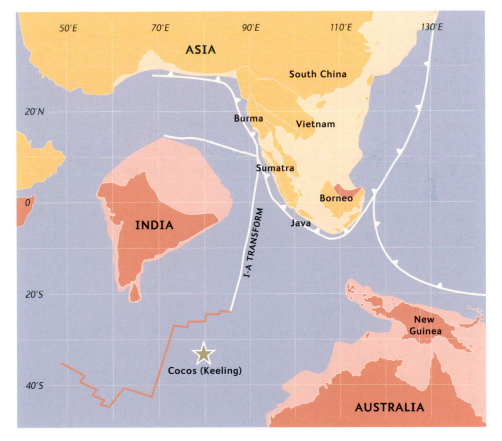

I think

Case must be that one generation then should be as many living as now. To do this & to have many species in same genus (as is) requires extinction.

Thus between A & B immens[e] gap of relation. C & B the finest gradation, B & D rather greater distinction Thus genera would be formed. — bearing relation

CHAPTER 6

~

The making of a theory

BY THE TIME THE *BEAGLE* REACHED the shores of England, thoughts of a country parsonage had long receded from Darwin's mind. He now aspired to be a gentleman scientist, and at the very least would publish his voluminous observations from the voyage and the theories he had developed. His geological work was uppermost in his mind, and from the Atlantic island of St Helena, three months from home, he wrote to Henslow asking if he or Sedgwick would propose him for membership of the Geological Society of London. Unknown to Darwin, his name was already becoming known in the scientific community. Some of Darwin's fossil mammals had been shown at a meeting of the British Association for the Advancement of Science in 1833 (see Chapter 2). Henslow and Sedgwick had also read extracts of Darwin's letters to the Geological Society and to the Cambridge Philosophical Society in 1835, and Henslow even had the extracts printed as a pamphlet for private circulation. A wider readership heard about it too, with the popular magazine *The Athenaeum* mentioning Darwin's *Megatherium* (ground sloth) discoveries and praising his geological work in the Andes.

Darwin was at first horrified to learn that his private letters had been made public in this way. But at the next port of call, Ascension Island, he received a letter from his sister Susan, quoting Professor Sedgwick who had written: 'He is doing admirably in S. America, and has already sent home a Collection above all

The page from Darwin's 1837 *Transmutation Notebook* in which he outlined his concept of evolution as a branching tree.

praise… He will have a great name among the Naturalists of Europe.' Henslow had written in a similar vein, 'rejoicing that you would soon return to reap the reward of your perseverance and take your position among the first Naturalist[s] of the day'. Darwin later recounted that on receiving Susan's letter, 'I clambered over the mountains of Ascension with a bounding step and made the volcanic rocks resound under my geological hammer!'

The *Beagle* docked at Falmouth in Cornwall on 2 October 1836. After visiting his family and rejoining the *Beagle* in London to unload the remainder of his belongings, Darwin took lodgings in Cambridge to consult with Henslow and, with the help of his *Beagle* servant Syms Covington, to unpack and sort his collections. There were also important people to meet – Sir Charles Lyell, whose *Principles of Geology* had been crucial to Darwin's work on the voyage, invited him to a tea party on 29th October 1836, where he also met for the first time the anatomist Richard Owen of the Royal College of Surgeons, where Darwin's fossil bones had been sent. Darwin subsequently visited Owen at the College on several occasions in December 1836, and invited him to undertake research on the fossils. Whereas William Clift, the curator, had received and prepared the specimens as they reached the College during in 1833 and 1834, it was Owen who was now the rising star of comparative anatomy, having been elevated from his initial position as Clift's assistant to his recent appointment as Hunterian Professor. While Owen's fame would undoubtedly have been assured by his later, seminal work on fossil reptiles, dinosaurs and much else besides, his study of Darwin's fossil mammals launched his career as a palaeontologist and did much to establish his reputation as the 'British Cuvier'. Later, in February 1838, Owen was delighted to receive the Geological Society's Wollaston Medal for his description of *Toxodon*, Darwin's most celebrated fossil find (see Chapter 2).

Darwin himself read his first scientific paper, on the uplift of South America (see Chapter 4), at the Geological Society in January 1837. Not only was he now a member of the Society, but in February of the same year was elected to its Council. In March he moved from Cambridge to London, to be at the hub of scientific life and to interact with the various specialists who would work on his collections. In the same month he presented his ideas on the links between volcanism, earthquakes and the slow elevation of mountain ranges to the Geological Society, leading to heated debate as some eminent members advocated a sudden, catastrophic origin. But the Society's President, in his annual address the following year, declared that 'Mr Darwin has presented this subject under an aspect which cannot but have the most powerful influence on the speculations concerning the history of our globe'.

Darwin became Secretary of the Geological Society in 1838, and in 1840 was invited to join the Council of another learned body, the Royal Geographical Society. However, neither administrative work nor the bustle of London life was ultimately suited to his temperament or to his scientific plans. In 1841 he stood down from both positions and the following year moved with his young family to the village of Downe in Kent, where he spent the rest of his life. But these years were intensely productive ones for Darwin. His first book, the *Journal of Researches*, appeared in 1839 (later editions were renamed the *Voyage of the Beagle*). Based on his diaries and with the addition of some scientific notes, it recounted his experiences of the voyage, becoming a best-seller and enhancing his reputation. German explorer and polymath Alexander von Humboldt, whose writings had first inspired Darwin in his travels, wrote to him to say that the book was 'one of the most remarkable works that, in the course of a long life, I

Darwin's study at Down House, showing the armchair in which he wrote most of his books, including *The Origin of Species*.

have had the pleasure to see published'. Darwin then produced, in succession, three volumes that gave detailed accounts of his geological discoveries from the voyage: *Coral Reefs* in 1842 (see Chapter 5), *Volcanic Islands* in 1844, and *Geological Observations on South America* in 1846, all of them including detailed accounts of his fossil finds and their significance to his conclusions. In January 1839 he had been elected a Fellow of the UK's leading scientific institution, the Royal Society, and in 1853 received its prestigious Royal Medal for these three works, as well as his subsequent researches on barnacles (see pp.204–5).

Evolution

At the same time as he was generating this prodigious output of books, Darwin was also quietly initiating his most profound contribution to science, as he laid down within the first few years of his return from the *Beagle*, all the key elements of his theory of evolution. Exactly when Darwin first became 'converted' to an evolutionary viewpoint has been much debated among scholars. According to one school of thought, it was not until early in 1837, when he received expert identification of his fossil mammals by Richard Owen, and of his Galápagos birds by the ornithologist John Gould. According to others, he was already having distinctly evolutionary thoughts about his observations during the course of the voyage, possibly quite early in the voyage. Whichever was the case, it is clear that his fossil finds, especially the fossil mammals, played a key role. He gave them maximum prominence in the famous first line of the *The Origin of Species*: 'When on board H.M.S. *Beagle* as naturalist, I was much struck with certain facts in the distribution of the organic beings inhabiting South America, *and in the geological relations of the present to the past inhabitants of that continent*' (italics added).

That he was struck by these facts while on board the *Beagle* itself has been regarded by some commentators as rhetorical flourish, but his other writings suggest that, for the fossil mammals at least, it was not an exaggeration. The huge carapace of *Glyptodon* seems to have made a particular impact; in his *Autobiography*, written for his children and presumably with no need for rhetoric, he wrote: 'During the voyage of the Beagle I had been *deeply impressed* by discovering in the Pampean Formation great fossil animals covered with armour like that of the existing armadillos', while in an 1864 letter to his German acolyte Ernst Haeckel he went further: 'I shall never forget my *astonishment* when I dug out a gigantic piece of armour like that of the armadillos' (italics added). Darwin was not a man to be astonished by something

A specimen of the small armadillo known as the pichi, *Zaedyus pichiy*, collected by Darwin near the bay of Bahia Blanca, close to Punta Alta where he unearthed remains of its extinct glyptodont relative.

without asking himself why it should be so, and in this case it was not just the wonder of the giant mammals, but their evident relationship to those still living in the same area, that was a key line of evidence for evolution. The example even made it into *The Origin of Species* where he wrote: 'In South America, a similar relationship is manifest, even to an uneducated eye, in the gigantic pieces of armour like those of the armadillo, found in several parts of La Plata' – making clear that even his 'uneducated eye' had noted the resemblance in the field.

Darwin admitted, however, that mammalian anatomy was not his strong point, leading some to suggest that only with Owen's identification of the fossils after the voyage did the significance of the extinct mammals dawn on him. However, it was not only the link from the glyptodont to the armadillo that he had deduced in the field. He had correctly recognized several of his giant sloth specimens as such, and knew of the work that had linked them to the smaller, living species (see Chapter 2). The key point was the congruence of geographical distributions between the extinct and living forms; both sloths and armadillos today have their core distributions in South America.

This relationship between the past and present occupants of a region was later formalized by Darwin as the 'law of succession of types'. The fossils that he had identified as rodents provided further examples. Taking them to be similar to the cavy or mara, he noted that the cavies 'are all proper to S America; and none have hitherto been found in a fossil state'. Although the bones later turned out not to be cavies (see pp.70–73), they are extinct relatives of other endemic South American mammals, so the principle remains. In an essay written in Chile in February 1835, Darwin wrote: 'I hope the Cavia of Bahia Blanca will be one more small instance of at least a relation of certain genera with certain districts of the earth. This correlation to my mind renders the gradual birth & death of species more probable.' The interpretation of the latter phrase as endorsing evolution should be treated with caution, as it was a direct quote from Lyell, then a trenchant opponent of evolution, who had meant only that species arose and became extinct sequentially, rather than in sudden acts of multiple creation or destruction. But in linking this 'gradual' procession to the geographical fidelity of mammalian groups, and in selecting the one example of all his finds where the fossil representative appeared to be only slightly different from the living one, it is hard to avoid the conclusion that Darwin was thinking in terms of an evolutionary replacement.

Lyell's *Principles of Geology*, Darwin's key text on the *Beagle*, provided additional fodder for these ruminations. In volume 2, Lyell had discussed theories of transmutation (evolution) at length, only to dismiss them, but Darwin must have read these chapters with intense interest. In volume 3, Lyell quoted a recent report by William Clift, describing fossil kangaroos and wombats from Australia, whose relationship to the living fauna of that continent was another clear example of the succession of types.

The living and the dead

In formulating his theory of evolution, Darwin linked the fossil evidence to key observations on living species. The fauna of the Galápagos Islands (visited in 1835) is celebrated in this regard, but before then (1833) he had already noted the fox-like mammal of the Falkland Islands, writing: 'Out of the four specimens brought home in the Beagle, three will be seen to be darker coloured, they come from the East Island. The fourth is smaller and rusty coloured, and is from the West Island… An accurate comparison of these specimens will be interesting.' In the Galápagos, it was the mockingbirds rather than the subsequently famous

The distribution of two closely related flightless birds – the common rhea to the north and Darwin's rhea to the south – was seen by Darwin as evidence for the evolutionary divergence of species.

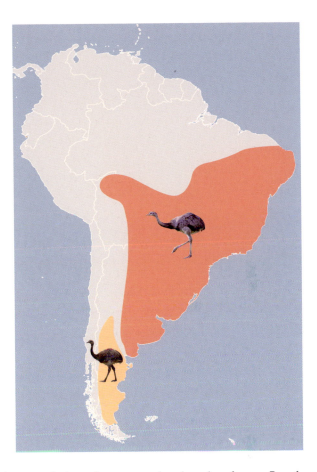

finches that caught his eye; he noted that they were closely related to a South American species, and that there were different 'varieties or distinct species' on the various islands of the group.

A slightly different case was that of the rheas, large flightless birds related to the ostrich. Darwin had seen rheas in the Pampas, and had heard that further south, in Patagonia, there was a smaller form known locally as the 'Avestruz petiso' (literally, short ostrich). He famously encountered one on his dinner plate one evening, and rushed to the *Beagle*'s kitchen to rescue its bones and skin before they were discarded. Later in the voyage he wrote: 'Whatever naturalists may say, I shall be convinced… that there are two species of rhea in S. America. In the plains of central Patagonia, I had several opportunities of seeing this ostrich: it unquestionably is a much smaller & darker coloured bird than the rhea.'

'Such facts ^would^ undermine the stability of species' – Darwin's first
recorded hint of evolution, written on board the *Beagle* in the
summer of 1836.

Revisiting his notes as the *Beagle* headed for home, Darwin wrote one of his most
famous passages. Having recounted the differences in the Falkland foxes and the
Galápagos birds, and noting that locals can tell from the shape of its shell from which
island any Galápagos tortoise came, he concludes: 'If there is the slightest foundation
for these remarks the zoology of archipelagos will be well worth examining; for such
facts would undermine the stability of species.' The last sentence was initially written
more boldly, without the word 'would', which was then inserted for caution.

Taking together all of Darwin's writings on his fossil and living specimens, it
seems highly likely that he was seriously entertaining evolutionary views during
the course of the voyage. When these thoughts began is difficult to say – some
would suggest not until 1836 when the above notes were written; others place
it conceivably as early as 1832 when he began to excavate his fossil mammals.
Certainly, the fossils were the first line of evidence he encountered and provided
the initial stimulus, as well as preparing his mind for what was to follow.

Confirmation (and elaboration) of his observations by the experts, in the months
following his return, were nonetheless significant in turning Darwin's evolutionary
speculations into convictions, and setting him on course for his life's work. In January
1837, Richard Owen broadly corroborated Darwin's suppositions about the fossil
mammals, and added one further apparent example of geographical continuity.
Owen declared that the *Macrauchenia* skeleton Darwin had unearthed at Port St
Julian in Argentina was a giant llama, thereby related to the living guanaco that
Darwin had seen in abundance on the Patagonian steppe. Sir Charles Lyell then

paraded *Macrauchenia*, with Darwin's other fossils and the Australian discoveries, in his February 1837 Presidential Address to the Geological Society, as evidence for the succession of types (but not, of course, for evolution).

At the same time, John Gould told Darwin that the his two rheas were indeed different species, and that the Galápagos mockingbirds, as well as an array of other birds that he identified as finches, were not only separate species particular to the different islands, but differed from their relatives on the mainland. For Darwin, the only explanation was that species, not just varieties, had actually evolved within the Galápagos group. Owen's and Gould's observations had confirmed his shipboard suspicions and clinched the argument. As historian of science Paul Brinkman has aptly put it, 'Darwin's adoption of transmutationism, like the elevation of the plains of Patagonia, happened gradually.'

The floodgates open

Very shortly afterwards, Darwin made a remarkable entry into his notebook, integrating into one sentence his evolutionary view of the birds and the fossils. 'The same kind of relation that common ostrich bears to petisse... extinct guanaco to recent: in former case position, in latter time.' By 'common ostrich to petisse' he meant the two species of rhea. By 'extinct guanaco to recent' he meant *Macrauchenia* and the living guanaco. The revolutionary idea was that change between two species in time is of the same kind as that observed in space at the present day. As Darwin later made clear, the evolutionary explanation of cases like the rheas is that the two species had descended from a common ancestor, one or both of them undergoing change in the process, probably within the areas they now occupy, over an unknown period of past time. In the mammals, fossil to recent, we witness that process in the time dimension. Darwin continued with an arresting phrase, asking whether, 'if one species does change into another', it would be sudden or gradual. At this early stage, he felt it must have been sudden; since the ranges of the two rheas abut, and there are no intermediates between them, he concluded that 'change not progressive; produced at one blow', and assumed that the same would have been true of the transition from the fossil to the living guanaco.

In July 1837, Darwin opened the first of a series of notebooks dedicated to the 'transmutation of species'. These notebooks, completed between the summer of 1837 and the end of 1839, are an exhilarating read for anyone with an interest in evolutionary biology. Having grasped the key organizing principle of evolution, ideas tumble out of Darwin's mind as he sees how the theory both explains diverse facts of natural history, and is in turn supported by them.

Darwin first pithily clarifies the significance of his key *Beagle* observations. Citing Galapágos tortoises, mockingbirds and the Falkland fox, evolution explains them: 'According to this view animals, on separate islands, ought to become different if kept long enough.' Similarly for the law of succession: 'Propagation [i.e. species descended one from another] explains why modern animals same type as extinct, which is law almost proved'. He further suggests why this law is most evident in terrestrial animals such as mammals: 'Geographical distribution of Mammalia more valuable than any other, because less easily transported'; by contrast, marine animals, or those that fly, will more quickly disperse around the world from their place of origin.

Darwin then moves beyond the law of succession and starts to consider evolution as a branching tree of life. This was an inevitable next step in a theory of species linked by common descent, but for Darwin the questions posed by his fossil mammals were clearly guiding his thinking, as his first notes on the tree-like relationships of species are linked to his fossil sloths and armadillos in relation to their living relatives. He wrote: 'We may look at Megatheria, armadillos and sloths as all offsprings of some still older type.' From the outset, Darwin was careful not to ascribe direct ancestry between known fossils and related living forms; the best we can do is to show that they have a more or less recent common ancestor. In the case of the sloths, German anatomists Christian Pander and Eduard d'Alton had proposed in 1821 that *Megatherium* was the ancestor of the living sloths. Richard Owen objected, pointing out that smaller sloths similar to those now living had been found as fossils contemporary with the giants; they were simply more resistant to the changed circumstances that led to the extinction of the larger ones. Darwin now brilliantly combined these perspectives – *Megatherium* has no direct descendants, but it and the living sloths (and the armadillos) were 'all offsprings of some still older type'. In modern parlance, they descended from a common ancestor.

Darwin then addresses the criticism, attributed to Cuvier, that if evolution were true we would have found intermediate forms between fossil species and those now living: 'Now according to my view', writes Darwin, 'in S. America parent of all armadillos might be brother to Megatherium – uncle now dead.' In other words, the ancestor ('parent') of living armadillos has not yet been found as a fossil: the best we have is an 'uncle' – *Megatherium*. He also illustrated, in his first sketch of an evolutionary tree, why living species may not form insensibly graded chains; instead they are linked by common ancestors that are extinct.

BELOW Darwin's first sketch of evolution as a branching tree. The three living species at the top may be traceable to a certain depth in the fossil record (solid lines), but the dotted ancestors are all extinct.

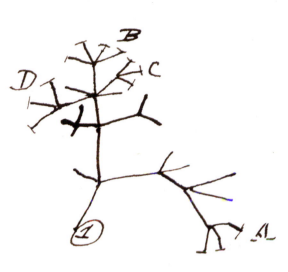

ABOVE Darwin's celebrated evolutionary tree, drawn a few pages after the first attempt (above). Species B and C are the most closely related, D more distant, A more distant still. (1) is their common ancestor.

Considering the fossil and living species in a tree-like framework helped Darwin to square the uncertainties in the relationships between his fossils and the living fauna. In the case of the sloths and armadillos, there was still confusion over whether the giant armadillo-like armour actually belonged to *Megatherium* (and hence pertained to the sloth family), or to a species more directly related to armadillos (see p.56). Rather than considering these forms as direct lines of descent, Darwin could recognize them as related to each other, in an admittedly uncertain order. When Owen's further work showed the fossil armour to belong not to *Megatherium* but to a group he named *Glyptodon*, the confusion resolved itself into two even clearer examples of succession: *Megatherium* and its kin to the living sloths, *Glyptodon* to the armadillo.

With *Macrauchenia* the situation was more awkward. Owen's assessment of it as a giant llama was short-lived – on more thorough examination he transferred it to the group including the rhinoceros (see Chapter 2) while still affirming its strong 'affinity' to the camel family (which includes the llama and guanaco). For him, in pre-evolutionary thinking, it formed a link between the two. A similar situation pertained to the giant mammal *Toxodon*, initially linked by Owen to the rodents, and hence a further example of succession, this time to the South American capybara, the largest living rodent. In his 1839 *Journal of Researches* Darwin was still declaring *Macrauchenia* and *Toxodon* to be allied to the guanaco and the capybara, respectively, but by the second edition of 1845 he was more circumspect – they bore relationships, but these were distant. In the case of *Macrauchenia*, Darwin's interpretation of Owen's link was that it might have descended from something close to the common ancestor of the camel and rhinoceros groups. Eventually, *Macrauchenia* dropped out of relationship with guanacos altogether, and *Toxodon* from the rodents – a somewhat ironic outcome, though not diminishing their significance in triggering Darwin's early exploration of evolutionary descent.

In the *Transmutation Notebooks* Darwin can be seen searching for a mechanism to account for the divergence of species implied by his evolutionary tree. His greatest discovery, natural selection, is generally taken to have been triggered by his reading of Thomas Malthus's *Essay on the Principle of Population* in September 1838, but he expressed a germ of the principle a year or so earlier, when he asked 'whether every animal produces in course of ages ten thousand varieties (influenced itself perhaps by circumstances) and those alone preserved which are well adapted.' However, the idea lay dormant until his reading of Malthus forcefully brought home to him how the overproduction of offspring, combined with checks to the number of survivors, had the power to drive change through many generations. This insight was not directly inspired by his study of fossils, although it was underpinned by the reality of evolution that the fossils, together with the evidence of modern distributions, had brought home to him.

The impact of Darwin's geological and fossil discoveries in leading him to an evolutionary viewpoint extended beyond the succession of types. That great changes can be wrought by a long series of incremental steps was evident from the gradual uplift of South America, as well as by the great depth of coral rock built up beneath oceanic atolls. By analogy major evolutionary transitions might be produced by the accumulation of small changes. That the Earth is immensely old – hundreds of millions of years at least – is also required if all of life's diversity has arisen by this process. This was demonstrated for Darwin by the great depth of geological deposition and erosion he had witnessed in South America, as well as by the scale of crustal uplift of the Andes.

Extinction

The extinction of species was, for Darwin, intimately connected to their evolution. If the environment could drive change within species, then it could also bring about their extinction. Even more importantly, a new, better adaped species might be expected to out-compete its progenitor, bringing about its extinction, and this was necessary if an approximately stable number of species were to be maintained in a given habitat.

Yet extinction, for Darwin, was a conundrum. Even in *The Origin of Species*, after 20 years of puzzling over the issue, he considered it 'the most gratuitous mystery'. It was the extinction of the large mammals, whose remains he had unearthed in South America, that lay at the root of his puzzlement. As he wrote in the *Journal of Researches*: 'It is impossible to reflect without the deepest astonishment, on the changed state of this continent. Formerly it must have swarmed with great monsters, like the southern parts of Africa, but now we find only the tapir, guanaco, armadillo, and capybara; mere pigmies compared to the antecedent races…. What then has exterminated so many living creatures?'

Darwin first dismissed the then prevalent view – promoted by Cuvier and accepted by his teachers, Henslow and Sedgwick – that species only became extinct en masse, during cataclysmic events that brought each geological period to a close. He had noted, firstly, that his extinct mammals occurred in beds filled with mollusc species that had survived to the present day (see p. 126); hence, as Lyell had pointed out before him, not all species had died out at the same time. To this he could add the example of his rodent fossils from Monte Hermoso, that he considered contemporary with the extinct giants; whereas the latter had left no descendants, the rodents had given rise to only slightly modified modern species.

Equally important against catastrophism, the geological context of his finds suggested that they had been deposited under normal, relatively calm conditions, not during a time of major environmental upheaval. Early in his collecting, at Punta Alta in the bay of Bahia Blanca, he had concluded that the beds, horizontally aligned with pebbles, shells and bones, were 'deposited by the action of tides… quietly'. This

A Late Pleistocene scene in Patagonia, with *Megatherium*
and horses (centre), glyptodont and sabre-toothed cats (left),
Macrauchenia (right), and *Notiomastodon* (background). Soon
after, all of these species were extinct.

has been described by the editors of Darwin's notebooks as a 'momentous entry' –
the deposits were not the relic of a sudden catastrophe but had formed in a river
or estuarine environment by processes similar to those of today. In a later essay,
Darwin enlarged on the idea - if there had been a major debacle, chaotic deposits,
with the inclusion of rocks and trees, would be expected; 'This is not the case in
S. America as far as I have seen.' The beds were the result of normal deposition
and the bones to 'a succession of deaths, after the ordinary course of nature'. To
these strongly Lyellian conclusions Darwin added his discovery of complete fossil
skeletons in the case of *Macrauchenia* and the ground sloth *Scelidotherium*, which
seemed to be lying at or close to the place they had died; a great upheaval would
surely have broken them up and dispersed the bones.

These conclusions might not have decided the issue of extinction were it not for another factor – Darwin's demonstration that the remains were geologically extremely recent, evidenced both on geological grounds and their association with recent shells. There had scarcely been time for, nor was there any evidence for, a great upheaval *subsequent* to the deposition of the bones. Therefore their extinction must have resulted from a much less dramatic cause, but what could that be? For Lyell, the answer seemed straightforward – since individual species are adapted to certain environments, and since we know that the Earth has undergone constant change, then those changes will periodically cause the extinction of species, and new species, better adapted to the changed circumstances, will take their place. Darwin's South American exemplar made it difficult for him to accept this simple solution. He believed that the deposits within which the bones occurred indicated little if any difference in the past environment compared to that of today. Moreover, it seemed unlikely that the gravel-strewn plains of Patagonia could ever have supported a more luxuriant vegetation than the present sparse scrubland. Even if there had been a certain change of climate, it cannot have been substantial – witness the unchanged composition of the marine fauna.

Darwin's conclusion that the past environment of Patagonia was as poorly vegetated as that of today brought fresh problems – how had large numbers of huge mammals managed to subsist there? The apparent solution came in June 1836 when the *Beagle* stopped off in South Africa for 18 days. There Darwin met with two local naturalists who described the abundance of large mammals living in relatively impoverished environments of the region. His notebook reads: 'Elephant lives on very wretched countries thinly covered by vegetation. Rhinoceros quite in deserts'. 'That large animals require a luxuriant vegetation, has been a general assumption', noted Darwin; 'I do not hesitate, however, to say that it is completely false.'

While these conclusions explained the former existence of large mammals in Patagonia, they made their extinction through changed environments even harder to accept. There was also the unique issue of his fossil horse teeth (see pp.60–61) – since reintroduced horses thrived so well in the present-day habitat, why did they go extinct in the first place? Finally, by this time it was becoming known that megafauna had died out, recently by geological standards, in many parts of the world – mammoths in Europe and Siberia, kangaroos in Australia, mastodons in North America. Darwin exclaimed in his notebook: 'It is a

Guanacos in the modern landscape of Patagonia. Darwin pondered
how the extinct giant mammals could have flourished in such a 'dry
and sterile country'.

wonderful fact, horse, elephant and mastodon dying out about same time in
such different quarters. Will Mr. Lyell say that some circumstance killed it over
a tract from Spain to South America? Never.'

The only serious contender for a global event sufficient to have exterminated
megafauna worldwide was Louis Agassiz's glacial hypothesis, first revealed in
1837. It had been proposed in catastrophist vein, a global glaciation wiping
out all life on Earth. This was certainly too much for Lyell, who refuted it with
examples where bones of megafauna had been found above supposedly glacial
deposits. Darwin agreed that the large mammals had survived the glacial period,
citing the lateness of his South American fossils. He added, presumably based
on the hardiness of the beast: 'Horse at least has not perished because too cold.'

Largely by default, Darwin turned briefly to the idea proposed by Italian geologist Giovanni Brocchi that species, like individuals, had a finite lifetime; as summarized by Lyell (who roundly rejected the idea), Brocchi proposed that the longevity of a species 'depends on a certain force of vitality, which, after a period, grows weaker and weaker…'. In his February 1835 essay, seeing no environmental cause for the extinction of the South American megafauna, Darwin flirted with the Brocchian concept, and even in an 1837 notebook after his return, wrote: 'Tempted to believe animals created for a definite time – not extinguished by change of circumstances.'

Soon after, however, Darwin recognized that such an idea was not biologically sustainable, accepting that in a general sense extinction must be 'a consequence… of non-adaptation of circumstances'; this was, indeed, almost implied by his theory of evolution by natural selection. Thus, 'extinction… will take place when conditions are unfavourable to numbers of animals, as in changing from warm to cold, damp to dry'. As far as the South American megafauna were concerned, he could only reflect on how little we understand the precise ecological conditions that determine the success, or otherwise, even of species alive today; hence we can 'argue with still less safety about either the life or death of any extinct kind'.

That humans might have played a part in the megafaunal extinctions appears not to have been considered by Darwin. Both Lyell and Owen had rejected evidence that humans had coexisted with the megafauna, arguing that people had appeared on Earth after their extinction. Lyell dramatically changed his mind in his 1863 book *Geological Evidences of the Antiquity of Man*, and while still emphasizing environmental causes, admitted that 'the growing power of man may have lent its aid as the destroying cause of many Post-Pliocene species'.

As for competition between species, Darwin's early writings doubted it as a plausible explanation: 'Will it be supposed that the armadillos have eaten out the Megatherium, the guanaco the camel?' In his mature view, however, expressed in *The Origin of Species*, Darwin came to see competition between species as a major driving force of extinction. Since natural selection is constantly producing better-adapted species, 'the consequent extinction of less-favoured forms almost inevitably follows'. Environmental change may also play a part; for example, the preferential extinction of larger species among the mammals may be due to their requiring a greater amount of food.

CLIMATIC AND HUMAN INFLUENCE ON EXTINCTIONS

Today, the causes of past extinctions are still debated, in much the same terms as those considered by Darwin and his contemporaries. In general, through the fossil record, environmental change is seen as the dominant factor, with direct competition playing a significant but lesser role. In the particular case of recent megafaunal extinctions, the principal contenders are ice-age climate change, and hunting by people. We now know that climate changes were highly complex, and not limited to times of actual glaciation, so Lyell's observation that many extinct megafauna survived beyond the glacial episode was correct. In South America, some 66 species of large mammal died out, mostly between about 15,000 and 8,000 years ago. Climate changes around this time led to the spread of forest and to changes in grassland habitats, which would have impacted the large herbivores. At the same time, humans had arrived in South America by 15,000 years ago, and there is clear archaeological evidence that they utilized species such as sloths and mastodons for products such as meat, skin, bone and tendon. Evidence for butchery, however, such as cut marks on bones, does not distinguish between hunting the living animal and scavenging a carcass; only a very few finds clearly indicate actual hunting, such as the recent discovery of a stone tool embedded in the skull of a young mastodon in southeast Brazil. The relative contributions of hunting and climate change to the extinction of the megafauna, as in Darwin's day, remain to be determined, but a synergistic effect of the two seems increasingly likely.

The origin of fossils

Darwin's conclusions about the extinction of the large mammals he had discovered were strongly influenced (as discussed on p.196), by his observation that they had been deposited 'quietly', under conditions similar to those of today. To underpin this conclusion he took great interest in the fate of animal remains. Lyell had discussed at length how dead animals and plants, or parts of them, get buried and might be preserved; such considerations were fundamental to his 'uniformitarian' approach to interpreting fossil strata. For Darwin, as for Lyell, it was also a strong argument against catastrophism. On the coast near Rio, he wrote, 'It most forcibly struck me' how 'a small change of level would immerse many terrestrial animals either in fresh or salt water deposits', producing strata like the Tertiary ones. At Montevideo, it was 'curious how very perfectly this dry sandy soil preserves insects... how easily would these have been preserved in any geological strata'. In the Falkland Islands he concluded, 'I believe the peat to be formed very slowly, from the grass and other plants now growing on the surface — I think so from seeing bones... lying on the grass, becoming partially enveloped'. Most dramatic of all, at Buenos Aires Darwin heard about the 'Gran Seco' (great drought) of 1827–1830, when large numbers of animals had perished through lack of food and water, including an estimated one million cattle. Eye-witnesses told him of thousands rushing into the river, 'whence exhausted by hunger, they were unable to ascend the muddy banks'; the river became 'full of putrid carcases', many of which floated downstream to the Plata estuary. After the drought came a great flood, 'Hence it is probable that some thousands of those skeletons were buried by the deposits of the very next year.' A geologist encountering such a deposit, Darwin surmised, might suppose some great debacle to have deposited them, yet they had accumulated according to 'the common order of things'.

Palaeontology after the *Beagle*

Although he barely did any more fossil-collecting after the *Beagle* voyage, Darwin continued to take a keen interest in fossils throughout his life, corresponding with palaeontologists and keeping up with the latest discoveries. The first significant addition resulted from the expedition of two Danes, Peter Lund and Peter Clausen, who from 1835 collected fossils from caves in Brazil, and published their findings in English in 1839. As Darwin enthused in his *Journal of Researches* the same year, their collection included 'extinct species of all the thirty-two genera, excepting four, of the terrestrial quadrupeds now inhabiting the provinces in which the caves occur... there are fossil ant-eaters, armadillos, tapirs, peccaries, guanacos, opossums, and numerous South American gnawers and monkeys'. It was striking confirmation of his law of succession of types and, privately, his evolutionary interpretation of it. Darwin also noted with satisfaction the finding of more complete remains of the species he had discovered. Skeletons of the ground sloths *Mylodon* and *Megatherium*, and further remains of his prize discovery, *Toxodon*, were sold to the British Museum in 1845 by the antiquary Pedro de Angelis, as well as fossils of the sabre-toothed cat *Machairodus*, 'a wonderful large carnivorous animal'.

Of more personal significance to Darwin, his friend from the *Beagle*, Bartholomew Sulivan, sent him lengthy accounts of the geology he saw during his further South American voyages, and even on occasion some fossils. In 1844 at Port Gallegos in southern Patagonia, Sulivan discovered a deposit rich in mammalian remains, including 'the whole back bone and ribs of an animal about the size of a small deer' and 'a piece of the shell of an armadillo quite perfect'. 'Of course they are for you', wrote Sulivan, 'but I suppose you will prefer their going with yours to the College of Surgeons.' Shortly after Sulivan's return to Britain in 1846, Darwin wrote to Owen that Sulivan was very anxious to have the fossils inspected by him, 'and I should be extremely glad to be present'. Sulivan's material formed the basis for Owen's new genus *Nesodon*, with two species, *N. imbricatus* and *N. sulivani*, although they are now regarded as one. The animals, about the size of a llama, were related to *Toxodon*, and Owen further remarked that 'The interval between Toxodon and Macrauchenia is evidently partly filled by the present remarkable genus.' Sulivan had presciently stated, from their geological position, 'I think they must be much older than any of the Bahia Blanca fossils' (where remains of *Toxodon* itself had been discovered by Darwin). The deposits are now known to be some 18-16 million years old, of similar age to invertebrate fossils collected by Darwin at Port St Julian and Santa Cruz, close to Port Gallegos (see pp.132–133).

Darwin's interest in new fossil discoveries was not limited to South America. His chief correspondent in this respect was the Scottish palaeontologist Hugh Falconer (1808–1865), with whom he exchanged cordial letters for 20 years. Falconer, with Proby Cautley, had excavated large quantities of fossils in the Siwalik Hills of northern India, presenting their collections to the British Museum and publishing their work in great monographs. In his notebook

Reconstruction of the fossil mammal discovered by Darwin's *Beagle* friend Bartholemew Sulivan on a later voyage. An early relative of *Toxodon*, it was named *Nesodon* by Richard Owen.

Darwin praised their 'account of wonderful fossils from India'. The finds were not without significance for Darwin; at one point he notes 'Falconer speaks of some fossil quite intermediate between Mastodon & Elephant' – a reference to the Siwalik mammal that Falconer later named *Stegodon*. When Falconer wrote with news of a fossil mammal from Brazil named *Mesotherium* ('middle mammal'), describing it as 'the common centre towards which all Mammalia got happily confounded', Darwin replied with glee: 'so intermediate a form is very glorious'.

The covert evolutionist

In May to June 1842, during a summer break at the family home at Maer in Warwickshire, Darwin wrote out a 16-page draft of his evolutionary ideas, heading it 'First Pencil Sketch of Species Theory'. Two years later, in 1844, he wrote a much longer version, running to 189 handwritten pages, in what was effectively a draft of *The Origin of Species*. On completion of the manuscript, he wrote a letter to his wife explaining his 'most solemn and last request' that she should see to its publication in the event of his death.

In the meantime, Darwin's published books avoided mention of evolution, although we can at times see him straining to reveal the true explanation for his observations. In the 1845 edition of his *Journal of Researches*, after detailing new evidence in support of his law of succession of types, he writes: 'This wonderful relationship in the same continent between the dead and the living, will, I do not doubt, hereafter throw more light on the appearance of organic beings on our earth, and their disappearance from it, than any other class of facts'. In the words of palaeobiologist Niles Eldredge, the law of succession had become a fig leaf for evolution. In another chapter Darwin invokes Lamarck, the most widely known proponent of evolution, as his mouthpiece. Writing of the common eye injuries sustained by the burrowing rodent the tuco-tuco, Darwin remarks that 'Lamarck would have been delighted with this fact… when speculating… on the *gradually acquired* blindness of the Aspalax' (another subterranean rodent; italics original). Covert evolutionary hints can also be seen in his comments on *Toxodon* in the same volume (see p. 76).

In 1846 Darwin embarked on eight years of intensive work on barnacles. It began with a study of some unusual living barnacles he had collected in Chile, but he realized that he needed to examine other species to put his find in context, and ended by producing a 1,000-page monograph on the entire group. In 1849

Darwin in 1849, aged 40, the year he began studying fossil barnacles.

he turned his attention to fossil barnacles, writing to all his palaeontological contacts in Britain and abroad for specimens and information. This resulted in another comprehensive study and a monograph in two volumes. Darwin's broader aim in this work was to become intimately familiar with one group of organisms, in the expectation of finding evidence in support of his evolutionary ideas. He was successful, especially in showing how series of living species showed stages in the modification of structures, illustrating plausible links in an evolutionary chain. He also found considerable variation *within* many species, a necessary prerequisite for his mechanism of natural selection, and for which most of his evidence up to that point had been derived from domesticated species. This applied to fossil barnacles as well, as Darwin noted that for many species it was 'necessary, on account of the variability of the characters, to possess several specimens' in order to properly identify the species' characteristic features.

The fossil barnacles, traced from the Jurassic Period to the present (around 200 million years of geological time as now understood), showed species appearing and disappearing in sequence. Darwin's published commentary at times reveals his secret evolutionary agenda, describing one of the more ancient species as the 'stem of the genealogical tree', and species leading 'with hardly a break' one to the other in their morphological progression. He also noted, for one fossil type found through a long series of strata (the Cretaceous stalked barnacle *Scalpellum arcuatum*), that 'there is a slight difference between the specimens from the upper and lower stages, which some authors might perhaps consider specific' – a coded suggestion that it wasn't a single species but that one species had 'perhaps' transformed into another.

Fossil barnacles, from Darwin's 1854 monograph. The shell comprises six overlapping plates, often found isolated (bottom row); Darwin used their detailed form to work out the relationships among species.

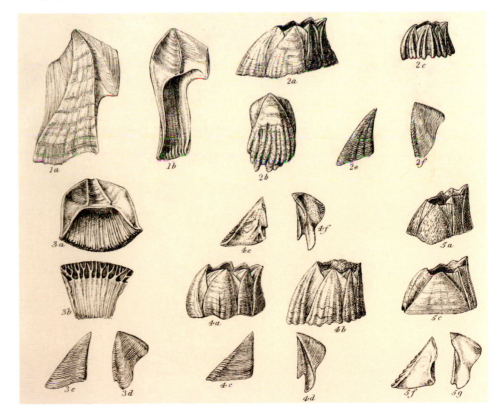

Fossils and *The Origin of Species*

In 1854, his barnacle work behind him, Darwin began work in earnest on his 'big species book', but was interrupted in 1858 by a letter from naturalist Alfred Russel Wallace, announcing a new proposal for species origins that was in essence identical to Darwin's theory of natural selection. Darwin and Wallace shared the honours with a joint presentation at the Linnean Society of London in July 1858, but it was Darwin who, with 20 years of accumulated evidence, wrote the book that gave the theory to the world. Contracting his planned big book into an 'abstract', and with his 1844 draft as a template, *The Origin of Species* was written in 15 months and published in November 1859.

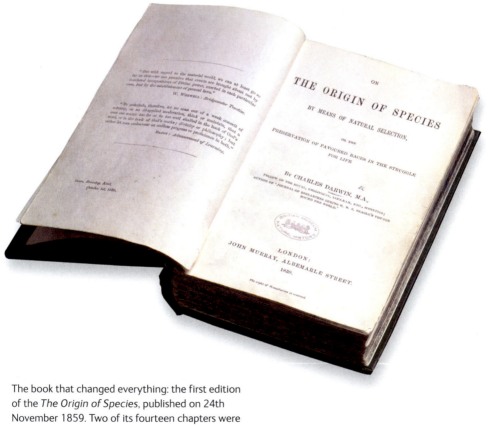

The book that changed everything: the first edition of the *The Origin of Species*, published on 24th November 1859. Two of its fourteen chapters were devoted to geology and fossils.

Two of the fourteen chapters of *The Origin of Species* are devoted to the fossil record. Darwin's account of geology and fossils has sometimes been portrayed as an apologia for the lack of fossil evidence for evolution, but it can be read in a much more positive vein, especially taking into account the examples that he added to successive editions, as new fossil discoveries came to light. In the first chapter, 'On the Imperfection of the Geological Record', Darwin admits that his theory predicts change in fossil species through thick rock sequences, and that demonstrated examples were very few. He explains that only a tiny proportion of living organisms are buried after death and that most of those will decay or dissolve away. Most rock sequences are subject to periodic halts during their formation, and to erosion thereafter, so the record is highly discontinuous. Darwin further astutely notes that new species are often likely to have arisen in a small part of their ancestor's range, and that many parts of the world are very poorly sampled for fossils, so it is unsurprising that transitional forms are rare. Following Lyell, he likens the geological record to a book of which many chapters are missing, most pages had been torn out, and on each page, only a few lines are preserved.

But Darwin goes on to show that, despite its imperfections, the fossil record speaks clearly in favour of evolution. Finely graded sequences showing species transformations may be rare, but at a slightly coarser resolution, comparing consecutive stages in a geological formation, we find that the embedded fossil species 'are far more closely allied to each other than are the species found in more widely separated formations'. By the final edition of *The Origin of Species* in 1872, Darwin could cite published accounts of change within species followed up a rock column, including ammonites from Russia and freshwater snails from Switzerland. Fossil shells from the late Tertiary, formerly believed to be the same species as those of today, had been shown to differ slightly from them.

The apparently sudden or late appearance of major groups of organisms had also been cited as evidence against evolution. Even in 1859, however, Darwin could quote examples that had been overturned by recent finds. For example, 'Cuvier used to urge that no monkey occurred in any Tertiary stratum; but now extinct species have been discovered in India, South America, and in Europe even as far back as the Eocene stage'. Darwin himself had been troubled by the 'sudden' appearance of non-stalked barnacles in the Tertiary Period; but no sooner had his barnacle monograph been published than one was found in earlier (Cretaceous) deposits in Belgium. By 1872 he could add that bony fishes, formerly supposed to appear suddenly in the Cretaceous, had been found in rocks up to nearly twice as old (Jurassic and Triassic). Not only did these facts argue against the late, sudden creation of major groups, but they showed that the evidence of the fossil record would only increase with continuing research.

In the next chapter, 'On the Geological Succession of Organic Beings', many additional lines of evidence are shown. Importantly, extinct and living species 'all fall into one grand natural system, and this fact is at once explained on the principle of descent'. It is also observed that 'the more ancient any form is, the more, as a general rule, it differs from living forms'. Finally, the law of succession of types could now be paraded in its true colours – it was the result of the early inhabitants of each continent leaving descendant species of the same general form, even if 'in some degree modified'. Moreover, whereas the living sloths and anteaters could not have descended directly from their giant relatives *Megatherium* and *Glyptodon*, there were now fossils from Brazil 'which are closely allied in size and in other characters to the species still living in South America; and some of these fossils may be the actual progenitors of living species'.

The years following the publication of *The Origin of Species* in 1859 saw the discovery of further key fossils showing stages in the origin of major groups of organisms, and these were eagerly incorporated by Darwin in later editions of the book. The most famous 'intermediate' fossil of all, *Archaeopteryx*, was discovered at Solnhofen, Germany in 1861. The first specimen was purchased by the British Museum and examined by Richard Owen, who presented his findings at a meeting of the Royal Society in 1862. *Archaeopteryx* had feathers like a bird, but claws on its wings and a long bony tail, like a reptile. Hugh Falconer was in the audience and wrote to Darwin: '… there has been this grand *Darwinian* case of the *Archaeopteryx* for you and me to have a long jaw about. Had the Solnhofen quarries been commissioned – by august command – to turn out a strange being à la Darwin, it could not have executed the behest more handsomely than in the *Archaeopteryx*'. Darwin replied immediately: '… I particularly wish to hear about the wondrous bird; the case has delighted me, because no group is so isolated as birds', and the next day he wrote to US geologist James Dana that *Archaeopteryx* was 'by far the greatest prodigy of recent times. It is a grand case for me…'. In *The Origin of Species*, the first mention of 'that strange bird, the Archaeopteryx' appeared in the fourth edition in 1866, but Darwin was here cautious, citing it mainly as an example of 'how little we as yet know of the former inhabitants of

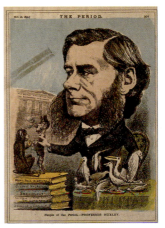

ABOVE AND BELOW Thomas Henry Huxley, Darwin's staunch defender, with the early bird *Archaeopteryx* (left) and the dinosaur *Compsognathus* (below), fossils discovered shortly after the publication of *The Origin of Species* that provided clues to the evolution of birds.

the world'. In the fifth (1869) edition, however, having consulted with Thomas Henry Huxley, now his authority on comparative anatomy in place of Owen, he added: 'Even the wide interval between birds and reptiles has been shown by Professor Huxley to be partially bridged over in the most unexpected manner by, on the one hand, the ostrich and extinct Archaeopteryx, and on the other hand, the Compsognathus, one of the Dinosaurians.' *Compsognathus*, a small running dinosaur of the theropod group, had also been found at Solnhofen, and Huxley had noted its bird-like features. He considered *Archaeopteryx* a bird rather than an intermediate, but its primitive features revealed that birds had evolved from reptiles. *Compsognathus*, contemporary with *Archaeopteryx*, was too late to be its actual ancestor, but was a survivor of the group that had given rise to birds. These conclusions remain essentially valid today, with some experts arguing that *Compsognathus* bore feathers.

By the sixth edition of *The Origin of Species* in 1872, there were further striking examples of fossil intermediates to exhibit. An extinct sea-cow, *Halitherium*, showed rudimentary hind limbs, unlike its living relatives and demonstrating their descent from a land-living ancestor. Fossil whales had been discovered too: *Squalodon* and *Zeuglodon* showed sharp, cusped teeth like other mammals and unlike the peg-like teeth of modern toothed whales; *Zeuglodon* also had hind-limb rudiments long enough to project from its body, in contrast to any living whale. Fossil horses (*Hipparion*) had been discovered with three toes in place of the single toe (hoof) of those of today: 'No one will deny', wrote Darwin, 'that the Hipparion is intermediate between the existing horse and certain older ungulate forms.'

In 1859 Darwin had concluded that the 'great leading facts in palaeontology seem to me simply to follow on the theory of descent with modification through natural selection'. By 1872 he was bolder: the 'great leading facts in palaeontology *agree admirably* with the theory of descent with modification through variation and natural selection' (italics added).

Darwin avoided mention of human evolution in *The Origin of Species* but addressed it in his 1871 book *The Descent of Man*. However, the fossil record relevant to human origins was at that time extremely paltry, and the book was mostly concerned with explaining modern human variation and adaptation in terms of natural and, especially, sexual selection. He suggested that human origins might lie in Africa, given that it was the home of gorillas and chimpanzees, but it was 'useless to speculate on this subject' as the only fossil ape then known, the Miocene *Dryopithecus*, had been found in Europe.

ABOVE AND RIGHT Hugh Falconer, Darwin's closest palaeontological colleague, with the human skull (right) he brought back from Gibraltar and took to Down House to show Darwin.

One of the first fossils to be recognized as that of an early human was a skull excavated in Gibraltar in 1848 and brought to England in 1864. In a letter to his botanist friend Joseph Hooker, Darwin recorded: 'Both Lyell & Falconer called on me & I was very glad to see them. F. brought me the wonderful Gibraltar skull.' Only much later was the Gibraltar skull recognized as that of a Neanderthal, the species that had been founded on a partial skeleton discovered in Germany in 1856. That skeleton showed an oval skull with a low, receding forehead and distinct brow-ridges above the eyes. Yet its brain capacity appeared as large as that of modern people. Huxley, in his *Evidence as to Man's Place in Nature* (1863), considered Neanderthal too close to modern humans to provide an intermediate form between humans and other apes. Darwin, in *The Descent of Man*, followed suit, while pointing out (correctly, as it turned out) that 'those regions which are the most likely to afford remains connecting man with some extinct ape-like creature, have not as yet been searched by geologists'. Those regions, Africa in particular, have now provided such remains in abundance.

After *The Origin of Species*

The initial response of scientists to *The Origin of Species* ranged from the jubilant to the enraged. At the Museum of Practical Geology in London, palaeontologist John Salter had prepared, within months of the book's publication, a display of fossil brachiopods 'arranged', wrote Darwin with satisfaction, 'after my diagram in the Origin, and it astonished me what a beautiful branching gradation he made by intercalating the varieties and species according to geological age'. On the other hand, John Phillips, President of the Geological Society at the time of publication of *The Origin of Species*, objected to Darwin's theory on the grounds that most species appear in the fossil record, persist through one or more formations, and then disappear, without evident change. Although Darwin did expect to see species changing in at least some fossil sequences, he did not consider examples of constancy fatal to his theory. In his early notebooks he had written 'My very theory requires each form to have lasted for its time', and in the fourth edition of *The Origin of Species* (1866) made clear, 'I do not suppose that the process… goes on continuously; it is far more probable that each form remains for long periods unaltered, and then again undergoes modification'.

Alfred Russel Wallace, co-discoverer of the principle of natural selection. He greatly admired Darwin's *The Origin of Species*.

Among Darwin's associates, those who were already evolutionists rejoiced in *The Origin of Species*. Wallace was gracious to a fault, proclaiming that Darwin's name should 'stand above that of every philosopher of ancient and modern times', and expressing relief that it had not fallen to him to give the theory to the world. Darwin's Edinburgh mentor Robert Grant, despite their having lost contact for 30 years, dedicated his 1861 book *Recent Zoology* to his former pupil, in 'admiration and approval'. Hooker and Huxley had at first been sceptical of evolution, but were bowled over by *The Origin of Species* and became Darwin's most outspoken champions. Henslow and Lyell, though reluctant to reject divine providence in nature, admired the brilliance of Darwin's arguments and helped to promote them. Lyell, in particular, gave an extended précis of Darwin's arguments in his 1863 *Geological Evidences of the Antiquity of Man*, broadly accepting them, though he suggested that the higher human faculties represented a 'leap' from the rest of the animal kingdom, and that its cause might not lie in 'the usual course of nature'. Sedgwick, like Henslow a man of the cloth as well as a geologist, was more critical, writing to Darwin that he had read the book 'with more pain than pleasure'. Robert FitzRoy, with whom Darwin had often discussed scientific matters on the *Beagle*, now expressed his regret at having given Darwin the opportunity to collect facts 'for such a shocking theory'.

The most complex and unfortunate case was Richard Owen. He had played a key role in his naming and describing of Darwin's fossil mammals, and the two had remained close through the 1830s and 1840s. Initially an opponent of evolution, from the mid-1840s Owen had expressed a belief in the origin of species by descent, probably stimulated by the evidence for regional succession of types. Owen's view was not entirely mechanistic, however; he believed in a Creator who was continuously operative through geological time and 'ordained' the birth of species. Moreover, Owen did not accept transmutation – the gradual modification of one species into another that underpinned Darwin's theory. He believed that new species arose in a sudden leap, due to major mutations in eggs, sperm or seeds. He was therefore bitterly opposed to the idea of natural selection, considering the theory to lack evidence and ignore the structural plan of living organisms. To make matters worse, in *The Origin of Species* Darwin had included Owen's name in a list of palaeontologists who had 'maintained the immutability of species'. Owen clearly made his feelings known on the matter, for in the second edition his name was absent from the list. Elsewhere in the book, Darwin had been complimentary, referring to him at one point as 'our great palaeontologist, Owen'. But Owen wrote a scathing review of *The Origin of Species*, and their relationship was at an end.

Conclusion

Darwin's own fossil finds, as naturalist on board the *Beagle*, had contributed significantly to the development of his ideas. The fact that he found them himself, in their geological context and underlying the modern landscape, was fundamental to their impact upon him. Their evident connection with the species he saw living in South America was also a key element in his conversion to evolution. The fossils provided his earliest examples of change through geological time, and were the raw materials from which he developed the idea of an evolutionary tree. They also led him to recognize the analogy between species' variation in time and space that has become a key element of subsequent evolutionary thinking. Later, Darwin was able to cite an ever-growing number of fossil discoveries as evidence in support of his theory. Aside from their evolutionary significance, Darwin made much use of his fossil finds in interpreting ancient environments, and they provided crucial evidence for his insights into continental uplift and the development of coral reefs. These also, in turn, gave strong support to Lyell's gradualistic view of the Earth's geological past, then not yet widely accepted.

Despite residual opposition, support for Darwin's ideas grew steadily. By the final edition of *The Origin of Species* in 1872, he could report that Lyell, formerly a creationist, 'now gives the support of his high authority to the opposite side; and most geologists and palaeontologists are much shaken in their former belief.' Darwin's theory revolutionized palaeontology, as fossils became the principal evidence that evolution had occurred. Relationships among organisms were no longer merely structural but genealogical, and reconstructing the tree of life became a major pursuit. Nonetheless, Darwin's insight that we can rarely identify a direct line of descent, and should focus instead on the nested relationships of living and fossil groups, was largely forgotten in the late 19th and early 20th centuries in the search for ancestral forms. Subsequently rediscovered, the emphasis now is on placing species within a branching tree, leading to an understanding of how successive changes have accummulated to produce major evolutionary transitions. In this endeavour, palaeontology is increasingly integrated with other areas of evolutionary biology such as ecology, genetics and embryology. The detail with which fossils now trace the origin of major groups such as birds and mammals, or smaller groups such as whales, would have astonished Darwin and his contemporaries. In rare cases, such as tiny organisms preserved in cores beneath the ocean floor, the geological sequence is complete enough to reveal gradual change and even the division of one species into two. Fossils are also widely used both in the reconstruction of past environments, and in understanding organisms' response to them, whether by evolution, shifts of range, or extinction. This revolution began the day Darwin stepped off the *Beagle* at its first port of call, and used the fossils he found in a cliff to conclude that the island of St Jago had recently risen from the sea.

SOURCES

Page numbers are in **bold**. Sources are listed by order of subject-matter on each page. References are to first printed editions except where stated. Letters are to or from CD except where stated. Quotations in the text have occasionally been corrected for idiosyncrasies of spelling and punctuation; in all cases reference to the original source is given below. Dates: 1.32 – January 1832; 1.1.32 – 1st January 1832.

ABBREVIATIONS

ACD The Autobiography of Charles Darwin (Barlow 1958); *AN 'A' Notebook*; *BBN Bahia Blanca Notebook*; *BD Beagle Diary* (Keynes 1988); *BN 'B' Notebook*; *BON Banda Oriental Notebook*; **CD** Charles Darwin; *CN 'C' Notebook*; *CPN Copiapó Notebook*; *CQN Coquimbo Notebook*; **CSD** Caroline Sarah Darwin; **DAR** Darwin manuscript numbers at Cambridge University Library (CUL-DAR); **DCP** Darwin Correspondence Project (Burkhardt et al. 1985–; https://www.darwinproject.ac.uk); *DN 'D' Notebook*; *DPN Despoblado Notebook*; **ECD** Emily Catherine Darwin (known as Catherine); *EN 'E' Notebook*; *GD Geological Diary* (Darwin 1832–6); *JR Journal and Remarks* (Darwin 1839), *JR45 Journal of Researches* (Darwin 1845); **JSH** John Stevens Henslow; *OS The Origin of Species* (Darwin 1859), *OS2–* later editions; *PDN Port Desire Notebook*; **RCS** Royal College of Surgeons; *RN 'Red' Notebook*; *RON Rio Notebook*; *SA Geological Observations on South America* (Darwin 1846), *SA76* 1876 edition; **SED** Susan Elizabeth Darwin; *SFN Santa Fe Notebook*; *SN Santiago Notebook*; *VI Volcanic Islands* (Darwin 1844); *VN Valparaiso Notebook*; *ZB Zoology of the Voyage of H.M.S. Beagle* (Owen 1838–40).

CHAPTER 1

7 CD quote: *ACD*, 76. **8** The survey: FitzRoy 1839; Browne 2008, xviii. FitzRoy quotes: Keynes 1988, xii; Fitzroy 1839, 18. **8–9** Choice of Darwin: letter from L Jenyns to F Darwin 1.5.82; FitzRoy 1839, 18–19; letter from JSH 24.8.31. **9** CD childhood: *ACD*, 22; Browne 2003/1, 29–33. **10** Edinburgh: Browne 2003/1, Ch. 3; *ACD*, 52; Herbert 2005, 33; Stott 2012. Cambridge: *ACD*, 62–4; Herbert 2005, 32; Kohn et al 2005. **10–12** Tenerife: Browne 2003/1, 135; Herbert 2005, 30; letters to CSD 28.4.31, JSH 11.7.31, WD Fox 1.8.31, from JSH 24.8.31. **12** Shropshire geology: letters to JSH 11.7.31, C Whitley 19.7.31. Welsh trip: Barrett 1974; Roberts 2001; Browne 2003/1, 142; Herbert 2005, 39–46. Accepted for *Beagle*: *ACD*, 54 & 60; Browne 2008, xvii. **13** CD quotes: *BD*, 6.1.32 & 16.1.32; **14** CD's cabin: letters to JSH 30.10.31 & ECD, 19.6.33; *BD*, 27.9.34; letter from Sulivan to Hooker (DAR107.42–7); Browne 2003/1, 170. **15–16** Shipmates: Keynes in *BD*, xxi–xxii & 61; *ACD*, 73–4; letter to R Fitzroy 20.2.40; Browne 2003/1, 202–210; Porter 1985, 985; *BD*, 29.8.33; FitzRoy 1839, 107. **17** CD & FitzRoy quotes: letter to ECD 29.7.34; FitzRoy 1839, 107. **18** Punta Alta: *BD*, 22.9.32 & footnote 1. Maldonado: Parodiz 1981, 53. **19** Map based on *BD*, 173, Winslow 1975 & Wesson 2017. **20** CD Santa Fe quote: *BD*, 6–11.10.33. Rosas & CD quote: *BD*, 10.20.33; *JR*, 87–89, 165; Parodiz 1981. Falklands: Armstrong 1992. Uruguay: *BD*, 3–4.10.33. CD quotes: letters to WD Fox 25.10.33 & CSD 20.9.33. **21** CD & FitzRoy quotes: *BD*, 2–4.10.32; FitzRoy 1839, 112, 126, 217; Cape Horn:

Chancellor & van Whye 2009, 61. **22** Andes crossing: *JR*, 390–4. **23** British connection: *BD*, 20.9.33, 26–27.9.34 & 15.3.35; *JR45*, 148. **24** CD quotes: *ACD*, 56; letter to SED 4.8.36. **25** Correspondence: letters to ECD 10.3.35 & WD Fox 9–12.8.35; *BD*, 25–26.4.33. **25** Packages home: letter to JSH 18.10.31; Porter 1985. **25–6** *Toxodon* package: letters to E Lumb 30.2.34, from E Lumb to JSH 2.5.34, from E Lumb 8.5.34, from C Hughes to W Clift 18.8.34 (RCS archive) & from CSD 29.12.34; Browne 2003/1, 226. Lima: letters to CSD 10–13.3.35, SED 23.4.35 & ECD 12.8.35. **26** *Beagle* writings: Burkhardt 2008; Chancellor n.d.; Porter 1985. Clift: letters from JSH 31.8.33 & 22.7.34, from CSD 28.3.34, to JSH 3.34 & CSD 8.8.34, from Clift to J Wedgwood 5.1.35 (RCS archive). **27** The specimens: *JR*, 599; letter to CSD 9.8.34; Rosen & Darrell 2010. **27–8** Disposal of collections: letters to JSH 9.9.31 & 30.10.36, CSD 9–12.8.34, from J Wedgwood to Clift 22.12.34 (RCS Archive); Burkhardt 2008, 39 footnote 3; Browne 2008, xix. **28** Hunterian Museum: RCS Board of Curators minutes, 19.12.36 (DCP-LETT-330); https://www.rcseng.ac.uk/museums-and-archives/hunterian-museum/about-us/history/. **29** Specialists: Porter 1985; Browne 2003/1, 451. CD quotes: letter to JSH 10–11.1832; *SA*, iv. **29–30** Rivalry: letters from d'Orbigny 14.2.45 & Sowerby 7.2.46; further letters at DAR43.1. **30** Early evolutionists: Stott 2012; Beccaloni & Smith 2015; Eldredge 2015, 75, 95; Browne 2003/1, 36–40, 83; Jameson 1827; *ACD*, 49. **31** Catastrophism and Lyell: Herbert 2005, 64, 156, 182–6; Pearson & Nicholas 2007; *ACD*, 77.

CHAPTER 2

35 Quotes: W Clift diary 8.1.33 (RCS Archive); letters from FW Hope 15.1.34, to CSD 6.4.34, to JSH 24.7.34. Evidence for evolution: *OS*, 1. **36** Map: based on Winslow 1975, Fig.3. Punta Alta: *JR*, 96. **37** *Megatherium* discovery: *BD*, 23.9.32. **38** *Megatherium* identification: letter to JSH 10–11.32. Further *Megatherium* skulls: *ZB*, 100–106. Luján skeleton: Simpson 1984; Argot 2008; Cuvier 1796 & 1812. **40** Parish's *Megatherium*: Parish 1839, 171–8; W Clift diaries 1832–1837 (RCS Archive). Cambridge meeting: letters from JSH 31.8.33, FW Hope 15.1.34, CSD 28.3.34 & J FitzRoy 24.8.33. **42** CD quotes: letters to JSH 3.34 & CSD 9–12.8.34. **42–3** Public interest: Rupke 2009; Toledano 2011; Dawson 2016; *GD*, 9–10.32. **43** Scientific accounts: Clift 1835; Buckland 1837; *ZB*, 100–106; Owen 1851–59; Dawson 2016, 79; Owen 1851–9, 823–8. Locomotion: Fariña et al. 2013; Pant et al. 2014; Blanco & Czerowonogora 2003. **44** Diet: Bargo & Vizcaíno 2008; Saarinen & Karme 2017; Bocherens et al. 2017. **45** CD quotes on *Mylodon*: letters to JSH 10–11.32 & 12.11.33; *BD*, 8.10.32. **46–7** '*Megalonx*' jaw: *ZB*, 99–100; Simpson 1984, 4–6. **47** Identification as *Mylodon*: R McAfee pers. comm., 2016. Parish's *Mylodon*: Owen 1842. **48–9** *Mylodon* biology: Moore 1978; Fariña et al. 2013; Barnett & Sylvester 2010. **49** Discovery of *Scelidotherium*: *GD*, 1833 (DAR32.74); *SA*, 84; *BBN*, 10a; letters to CSD 13.11.33 & JSH 3.34; *ZB*, 73. **50–1** Discovery of *Glossotherium*: *ZB*, 57–63; *SA*, 92; *JR*, 181; letter to T Reeks (DCP-LETT-823). *Glossotherium* & *Scelidotherium* biology: Bargo & Vizcaíno 2008; Fariña et al. 2013. **52** Sloth distributions: Fariña et al 2013, 213–6; Varela & Fariña 2016. **52–3** Sloth evolution: Pant et al. 2014; Buckley et al. 2015; Slater et al. 2016. **53** Punta Alta glyptodonts: *RON*, 64b; *BD*, 15.9.32; *GD*, 9–10.1832 (DAR32.65–6); *ZB*, 107. **54–5** Further Glyptodont finds: *GD*, 1833 (DAR32.74); letter to CSD 20.9.33; *JR*, 153 & 181; *JR45*, 130; *SFN*, 33a; *BON*, 36; *GD* 1833 (DAR33.258 verso). **55–8** Identity of carapace: letter to JSH 26.10–24.11.32; Falkner 1774; Owen 1841 (citing Laurillard, Pentland &

d'Alton); Clift 1832–7 & 1835; letter to CSD 20.9.33; *GD*, 1833 (DAR33.252–259, 270); Brinkman 2010; Darwin 1844 (in F. Darwin 1909); Herbert 2005, 303–8; DAR205.9; *SA*, 78 & 84. **58** Glyptodont biology: Delsuc et al. 2016; *BON*, 36; *GD*, 1833 (DAR33.258 verso); Alexander et al 1999; Bocherens et al. 2017. **58–60** Glyptodont taxonomy: Delsuc et al. 2016; *SFN*, 30a; *JR*, 181. Identification of *Neosclerocalyptus*: Alfredo Zurita & Fredy Carlini, pers. comms., 2016. **59–60** Biology of *Neosclerocalyptus* & *Glyptodon*: Zurita et al. 2008, 2010, 2011; Vizcaíno et al 2010, 2011; Fariña et al 2013, 231–2; Saarinen & Karma 2017; França et al 2015. **60–1** Horse teeth: *GD*, 1833 (DAR33.252–3); *JR*, 96 (footnote) & 149–151; *ZB*, 108–109; Simpson 1984, 28; Winslow, 1975; *OS*, 318–9. **62–3** *Equus neogeus*: Alberdi et al. 1995; Fariña et al. 2013; Prado et al. 2011; Prado & Alberdi 2014; Orlando et al. 2008. **63** Rio Negro flint: *BBN*, 27a ('Churichol' was probably Choele Choel); *BD*, 4–7.9.33. **63–4** Discovery of mastodons: *BD*, 1.10.33; *JR*, 147 (the Carcarañá is termed the Tercero); *SA*, 87–8; *GD*, DAR33.255; letter to JSH 12.11.33. **65–7** Identification of mastodon: Griffith 1827–35; Cuvier 1806; Simpson 1984; Owen 1845; map modified from Mothé et al. 2016. **66–8** Gomphothere evolution: Osborn 1936; Ferretti 2011; Lucas 2013; Dantas et al 2013. **68** Gomphothere biology: Larramendi 2016; Mothé et al. 2012; Asevedo et al. 2012. **68–9** Monte Hermoso: Deschamps et al. 2012; Fitzroy 1839, 112; *BD*, 19.10.32. **69–70** Fossils & geology: Zárate & Folguera 2009; Quattrocchio et al. 2009; *GD*, 10.32 (DAR32.69–72); *SA*, 81–4. **70–1** Foot bones: *GD*, 9–10.32 (DAR32.70–71 verso; following Cuvier, CD erroneously considered the mara a form of agouti); *ZB*, 109–110; Eldredge 2015; letter to JSH 3.34. **71–2** *Ctenomys* & capybara: Owen 1845, 36; *JR*, 59–60, 97. **73** Identification of *Phugatherium*: D. Verzi, pers. comm. 2017. **73–4** Biology of *Phugatherium*: Deschamps et al. 2012; Fernicola et al. 2009; Vucetich et al. 2013; *SA*, 82. **74** *Actenomys*: Morgan & Verzi 2011; Fernández et al. 2000, 74; *Paedotherium*: identification by Marcos Ercoli, pers. comm. 2017; Elissamburu 2004. **76** *Toxodon* skull:

Winslow 1975; *JR*, 180–1; *BON*, 33; letter to JSH 3.34. **77–8** 'Giant rodent' teeth: *GD*, 9–10.32 (DAR32.64); letters to JSH 26.10–24.11.32 & 12.11.33, to C Lyell 30.7.37; Owen 1845, 133; *JR*, 96, 146; *SA*, 84, 88, 90; *SFN*, 32–33a; Owen *ZB*, 16–35. **78–82** Interpretation of *Toxodon*: *ZB*, 16–35; *JR45*, 82; Owen 1837; Minutes, 19.4.37 (Geological Society archive); *JR*, 180–1; Brinkman 2010; letter to CSD 9.11.36. **81–2** Biology of *Toxodon*: Ameghino 1889; Bond 1999; França et al 2015; Fariña et al 2013, 203 & 281; Shockey 2001. Notoungulata: Owen 1853; *SA*, 89; Bond 1999. **83** Discovery of *Macrauchenia*: *GD*, 10–18.1.34; *SA*, 95; letters to JSH 3.34 & ECD 6.4.34; *JR*, 208. **83–84** The 'gigantic llama': Wilson 1972, 437; letter from C Lyell 13.2.37; Minutes, 17.2.37 (Geological Society archive); Lyell 1833–38b; *RN*, 130; Rachootin 1985. **84–88** *Macrauchenia* bones & name: *ZB*, 35–56; letter to R Owen 28.12.37; Owen 1840–45, 602–3; letter from G Waterhouse 30.3.46 (DCP-LETT-968 & footnote 2). *Macrauchenia* biology: Fariña et al. 2005 & 2013; Bond 1999. **90–1** *Macrauchenia* & *Toxodon* relationships: *OS*6, 125, 151 & 386; Welker et al 2015; Owen 1853; Cope 1891; Agnolin & Chimento 2011.

CHAPTER 3

93 Paraná wood: *GD*, 3–11.33 (DAR33.251); *SA*, 89; Iriondo & Krohling 2009. **93–4** Petrification: Anderson 2009; Kenrick & Davis 2004; *PDN*, 88–90. **95** Wood identification & microscopy: Kenrick & Davis 2004; Falcon-Lang 2012; Falcon-Lang & Digrius 2014. CD identifications & dating: *SA*, Ch. 5. **96–7** Santa Cruz & Chile: *SA*, 115; *VN* 82a; *GD*, 1.35 & 11.34 (DAR35.310 & 292); *PDN*, 88. **97** Discovery at Agua de la Zorra: *JR*, 405–7; *SFN*, 178a–180a; *GD* , 4.35 (DAR36.517–523); letter to JSH 18.4.35. **99** R Brown: Porter 1985; letter to L Jenyns 10.4.37; *JR*, 406. Modern reconstruction: Rößler et al. 2014; Brea et al. 2009; Brea 1997. **100** CD's interpretation: *JR*, 406; *GD*, 4.35; *SA*, 202–3. Modern interpretation: Poma et al. 2009. **101** Age & landscape: *SA*, 201–3; *BD*, 4–5.4.35. **102** Chilean logs: *CQN*, 36; *SA*, 208; *CPN*, 76–81;

GD, 6.35 (DAR37.618); *JR45*, 353; Chancellor & van Whye 2009, 488. **103** Illawara: Thomas 2009. **104** Tasmania: Thomas 2009; Philippe et al. 1998. Bald Head: *VI*, 144–7; *GD*, 3.36 (DAR38.858–863); *JR*, 537. **105** Chilean lignite: *GD*, 11.34 (DAR35.291); *PDN*, 81–2; *BD*, 6.3.35; Mishra n.d. **106** Coal from NZ & Australia: *GD*, 12.35 & 1.36 (DAR37.805–6 & 38.822); Hutton 2009. **107–8** Beech leaves: *GD*, 2–3.34 (DAR34.166); *SA*, 117–8; Heads 2006. **108–111** *Glossopteris*: *VI*, 130. **109** Drifting continents: Kious & Tilling 1996; Knapp et al. 2005. **111** Tasmanian travertine: Thomas 2009; *JR*, 535; *SA76*, 157–8; Jordan & Hill 2002.

CHAPTER 4

Note: Scientific names of species in this chapter are those currently accepted and often differ from names used by Darwin and his colleagues. Updated names are based on the cited publications, www.malacalog.org, http://fossilworks.org, www.marinespecies.org, http://www.bryozoa.net/famsys.html, and recent specialist identification of museum specimens.

113 St Jago & rhodoliths: Pearson & Nicholas 2007; Johnson et al. 2012; *VI*, 4 & 153–8. **115** Uplift vs subsidence: Herbert 2005, 154–5; *GD*, 1–2.32 & 5.34 (DAR32.23 verso & 34.40 verso); *SA*, 15. **115–6** St Joseph's, Port Desire & Port St Julian: *GD*, 1833–4 (DAR33.224 & 238; 34.40–60). **116** Santa Cruz: *GD*, 4–5.34 (DAR34.105–6 & 144–5); *SA*, 8–9. **117** CD's interpretation: *GD*, 1833–4 (DAR33.224 & 238; 34.40–60); Chancellor, n.d. **117–8** Chile: *VI*, 233–5; *GD*, 7.34 & 3.35 (DAR35.222 verso & 369–70); *CQN*, 8 & 15. **118** Shells as food: *SA*, 29; *GD*, 11.34, 2.35 & 7.35 (DAR35.296–7, 36.420 & 37.711–5); *SA*, 32. **120** Terrace formation: *GD*, 1833, 3.34, 7.34 & 7.35 (DAR34.21, 36.419, 35.213 & 37.693); FitzRoy 1839, 412–4. **120–1** Gradual uplift: *JR*, 203–8; *SA*, 14–18; letter to JSH 28.10.34; Herbert 2005, 160–6. **121–2** St Helena: *GD*, 7.36 (DAR38.920–935); *JR*, 582; *VI*, 88–90 & 153–8; Groombridge 1992; IUCN 2017. **123–5** Punta Alta: *GD*, 9.32 &

1833 (DAR32.64 & 74); *SA*, 83–4; Keynes 2000, xi–xvii. Identification of bryozoan: C. Sendino & P. Taylor, pers. comm. 2017. *Macrauchenia*: *SA*, 95–6. **126** Extinction: *SA*, 86 & 95; *JR*, 97. **126–7** Microfossils: Ehrenberg 1845, 143–8; *SA*, 85 & 88; Dumitrica 2007. **127–9** Patagonian terraces: Rostami et al. 2000; Schellman & Radtke 2010; Pedoja et al. 2011. **129–130** Pampean Fm & Punta Alta fossils: Zárate & Folguera 2009; *SA*, 82–86; Farinati 1985; *BD*, 24.12.34; Williams 2017. **130–2** Tertiary shells: *JR*, 201; *SA*, 118–9; *GD*, 4.33 & 1.34 (DAR33.223–226, 245–248, & 34.7–9); Lambert & Jeannet 1928; Pick 2004; Parras & Griffin 2009. **132** Tertiary microfossils: *SA*, 111; Ehrenberg 1845; figure from Ehrenberg 1854, Pl. XXII. **133–4** Bajada shells: *SFN*, 26a; *SA*, 89. **134–5** Summary of Tertiary shells: Griffin & Nielsen 2008; Casadío & Griffin 2009. From CD's collections, Sowerby & d'Orbigny identified 36 species (mostly molluscs) from Patagonia & 59 from Chile, including 55 new species of which 45 remain valid. Map based on Gross et al. 2015. Age of shells: letter to JSH 28.10.34; Lyell 1830–33, vol. 3; Herbert 2005, 63; Darwin 1834, 95a verso & 148b verso. **135** Navidad: *BD*, 262. **136–7** Sharks: *JR*, 423; *GD*, 1833 & 1.34 (DAR33.247 & 250); *SN*, 64. **138** Age of Tertiary deposits: Parras & Griffin 2009; Griffin & Nielsen 2008; Casadío & Griffin 2009; Iriondo & Krohling 2009; Encinas et al. 2014 & refs therein. **139** Subsidence: *SA*, 137. **139–141** Mt Tarn & Port Famine: *BD*, 2–6.2.34; *GD*, 3–6.2.34 & 1–7.6.34 (DAR34.125–8 & 153–6); **141–2** Ammonites: *SA*, 151–2 & 265; letter from E Forbes 7.8.46; Viens 2014; Kennedy & Henderson 1992; Olivero et al. 2009. **142** Osorno eruption: *BD*, 26.11.34; *JR*, 336, 356. **142–4** Concepción earthquake: *BD*, 20.2.35 & 5.3.35; FitzRoy 1839, 415; letter to JSH 10.3.35; Darwin 1840; Herbert 2005, 217–232. **144** Piuquenes Pass: *JR*, 390–4 & 415; *BD*, 20.2.35. **144–5** Piuquenes fossils: *GD*, 3–4.35 (DAR36.479); letters to JSH 18.4.35 & SED 23.4.35; *SA*, 181; Aguirre-Urreta & Vennari 2009. **146–7** Tomé nautiloid & ammonite: FitzRoy 1839, Ch. 19; *GD*, 3.35 (DAR35.358 verso); *SA*, 126–7; letters from E Forbes

7.8.46 & A d'Orbigny 14.2.45 (DCP-LETT-829 & editors' notes); Nielsen & Salazar 2011. **147–8** Arqueros rudists: *SA*, 212; D'Orbigny 1842, 107; Masse et al. 2015 (identification as *Jerjesia chilensis*); P. Skelton pers. comm. 2017 (identification as Monopleuridae indet.). Engraving (p. 147) *Hippurites toucasianus*; map (p. 148) based on Masse et al. 2015 (cf. *SA*, 61). **148–150** R. Claro to Despoblado: *GD*, 5–7.35 (DAR36.587 & 37.615); *SA*, 215–7, 223–4 & 265–8; *CPN*, 50; *DPN*, 3b & 15b; *JR*, 435. **151** Peru: *GD*, 13–14.7.35 (DAR37.678-9). **151–2** Uplift of the Andes: *CPN*, 71–2; *SA*, 242; Charrier et al. 2006; Darwin 1840; Herbert 2005, 218–228; Pedoja et al. 2011. **152–5** Discovery of Falkland fossils: Chancellor & van Whye 2009, 94; *BD*, 3.3.33 & 10-17.3.33; *GD*, 1833–4 (DAR32.125–7, 33.165–7 & 217–222); letter to JSH 4.11.33; Stone & Rushton 2012. **155–6** Murchison & Sowerby: *RN*, 142–144; *JR*, 253; Herbert 2005, 425. **156** Brachiopods & crinoids: Morris & Sharpe 1846; Stone & Rushton 2013. Trilobite: Stone & Rushton 2013; *JR*, 253; Aldiss & Edwards 1999. Identification as *Bainella*: G. Edgecombe, pers. comm. 2017. **156–9** 'Corals': *GD*, 1834 (DAR33.165–7); Stone et al. 2015. **158–9** Interpretation of Falkland fossils: letter to CSD 30.3.33; Murchison 1839; Darwin 1846; Morris & Sharpe 1846; *GD*, 1834 (DAR33.167); *JR*, 253; Armstrong 1992; Stone & Rushton 2012. **159** Tasmanian fossils: *VI*, 138–9 & 158–169; *GD*, 2.36 (DAR38.845). **159–161** Interpretation: Banks 1971 & Z. Hughes, pers. comm. 2017 (brachiopods *Spirifer, Fusispirifer, Neospirifer, Strophalosia, Sulciplica* & *Ingelarella*; gastropod *Peruvispira*); Wyse Jackson et al. 2011 & C. Sendino, pers. comm. 2017 (bryozoans *Fenestella, Protoretepora, Hemitrypa* & *Stenopora*); Morris 1845.

CHAPTER 5

163–4 Questions about atolls: Stoddart 1976. **164–5** Darwin's critique & theory: letter to CSD 29.4.36; Darwin 1835a & 1842, 88–118; *ACD*, 98; *SN*, 95–7. **165** Diagrams: modified from Darwin

1842; <u>Galápagos</u>: *VI*, 114–5; *GD*, 792; Herbert 2005, 170; Glynn et al. 2015. **166–7** <u>Pacific islands</u>: *BD*, 13.11.35, 23.11.35 & 3.12.35; *JR45*, 402–3; Darwin 1835a; **167–8** <u>Work on Cocos (Keeling)</u>: FitzRoy 1839, 33 & 38; Armstrong 1991. **168–171** <u>Keeling corals</u>: Darwin 1836, 1842; Rosen 1982; Rosen & Darrell 2010, 2011, who rediscovered Darwin's coral specimens at the NHM and have related them to his hand-written account of their positions on the atoll. **172** <u>Origin of material</u>: *BD*, 4.4.36; Darwin 1836; Perry et al. 2015. **173** <u>Building the atoll</u>: Darwin 1836, 9 verso; Stoddart 1995. **174** <u>Subsidence</u>: *JR*, 560; Li & Han 2015; Armstrong 1991. **174–5** <u>Soundings</u>: Darwin 1836, 20–27; Darwin 1842, 7–9, 72; Quoy & Gaimard 1824; Sponsel 2016. **175–6** <u>Mauritius</u>: *GD*, 5.36 (DAR38.885–897; specimen 3622 survives at the Sedgwick Museum, Cambridge); *BD*, 5.6.36; *VI*, 28; Herbert 2005, 239–240. **176–7** <u>The global picture</u>: Darwin 1842, 18 & Ch. 6; Darwin 1836, 6 verso. **177–9** <u>Theory put to the test</u>: Judd 1890, 5; Rosen & Darrell 2010; Stoddart 1976; Daly 1915; Dobbs 2005; letter to A Agassiz 5.5.81; Bonney 1904; Steers & Stoddart 1977; plaque and photo by B. Rosen. **179–180** <u>Ice-age effects</u>: Woodroffe 2005; Grigg 2011; Woodroffe et al. 1990, 1991; Woodroffe & MacLean 1994. **180–81** <u>Plate tectonics</u>: Rosen 1982; Scott & Rotondo 1983; map based on Hoernle et al. 2011 & Hall 2012.

CHAPTER 6

183 <u>Plans to be a scientist</u>: letters to CSD 29.4.36 & JSH 9.7.36. **183–4** <u>BAAS display</u>: letter from JSH 31.8.33. <u>Henslow's & Sedgwick's readings</u>: Browne 2003/1, 335–6; see also Lyell 1833–38a, 367–8. <u>The Athenaeum</u>: Anon. 1835. <u>The pamphlet</u>: Darwin 1835b; see also Burkhardt 2008, 362, note 1. <u>Letters</u>: to ECD 3.6.36; from SED 22.11.35 & CSD 29.12.35; *ACD*, 81–2. **184** <u>Owen</u>: Browne 2003/1, 348–9; Desmond & Moore 1991, 201–2 & 235. **184–5** <u>Learned societies</u>: Herbert 2005, 89; Browne 2003/1, 428; Desmond &

Moore 1991, 279; https://royalsociety.org/grants-schemes-awards/awards/premier-awards/. **185–6** <u>Books</u>: Darwin 1839, 1842, 1844, 1846; Browne 2003/1, 414–7; letter to John Washington 14.10.39 & editor's note 4 (DCP-LETT-537). **186** '<u>Conversion</u>' to evolution I: Sulloway 1982; Browne 2008, xxiv; Brinkman 2010; Eldredge 2015. **186–7** <u>Glyptodon</u>: *ACD*, 118; letter to E Haeckel 8–10.64; *OS*, 339. **188** <u>Law of succession of types</u>: *JR*, 210; *GD*, 10.32 (DAR32.71); Darwin 1835c, 2 recto; Lyell 1830–33, vol. 3, 33 & 143–4; Eldredge 2015. **188–9** <u>Foxes, mockingbirds & rheas</u>: Darwin 1832–3, 20; Darwin 1836–7, 262; Keynes 1988, 212; Keynes 2000, 189 & 298. **190–1** '<u>Conversion</u>' to evolution II: Judd 1911; Barlow 1963, 207; Eldredge 2015; Herbert 2005, 311; Chancellor n.d.; Lyell 1833–38b, 510–511; Browne 2003/1, 359–361; Brinkman 2010, 394. **191** <u>Guanacos & rheas</u>: *RN*, 127 & 130; Herbert 2005, 320–1; *BN*, 16: 'I look at two ostriches as strong argument for such change – as we see them in space, so might they in time'; Rachootin 1985. **191–2** <u>Transmutation notebooks</u>: Barrett et al. 1987, 6; *BN*, 7, 14 & 81; *OS*, 323–6. **192** <u>Evolutionary trees</u>: *BN*, 19–20 & 53–4; tree diagrams from *BN*, 26 & 36; Pander & D'Alton 1821; Owen 1851–9, 823–8. **193–4** <u>Fossil & living fauna</u>: *JR*, 180–1 & 209; *JR45*, 172–3; *CN*, 201: 'My theory drives me to say that there can be no animal at present time having an intermediate affinity between two classes – there may be some descendent of some intermediate link'; *OS6*, 302 & 378–9. **194** <u>Natural selection</u>: Kohn in Barrett et al. 1987, 167–8; *BN*, 80; *DN*, 153. **194–5** <u>Geology & evolution</u>: Wesson 2017, 260–1; *OS*, 282. **195** <u>The puzzle of extinction</u>: *EN*, 43; *OS*, 318–320; *JR*, 201–212. **195–6** <u>Against catastrophism</u>: *GD*, 9.32 (DAR32.71 verso); *RON*, 64b–65b; Chancellor & van Whye 2009, 36; Darwin 1835c. **197–8** <u>Extinction of megafauna</u>: Lyell 1830–33, vol. 2; Darwin 1835c; *RN*, 63, 129 & 134e; *JR*, 98–102 & 208–212. **198** <u>Glacial hypothesis</u>: Grayson 1984; Lyell 1844; *SA*, 96–7; *RN*, 113e. **199** <u>Expiry date or adaptation</u>: Lyell 1830–33, vol. 2, 128; *RN*, 129; *BN*, 46; *DN*, 37; *JR*, 211. <u>Human impact</u>

& competition: Grayson 1984; Lyell, 1863; *AN*, 9; *OS*, 317; *OS3*, 346. **200** <u>Climatic & human influence</u>: Benton 1987; Barnosky & Lindsey 2010; Prado et al. 2015; Barnosky et al. 2016; Metcalf et al. 2016; Mothé et al. 2017. **200–1** <u>The origin of fossils</u>: Lyell 1830–33, vol. 2; *GD*, 4–6.32, 9–10.32 & 3.33 (DAR32.53, DAR32.67 verso, DAR32.132 verso & DAR33.260–1); Iriondo & Krohling 2009. **201–2** <u>New fossils from S America</u>: Simpson 1984; *SA*, 106 & 117; letter from G Waterhouse 30.3.46. <u>Nesodon</u>: letters from B Sulivan 13.1–12.2.45 & 4.7.45; Brinkman 2003; letter to R Owen, 21.6.46; Owen 1846 & 1853, 309. **202–3** <u>Falconer</u>: *BN*, 126e; DAR205.9.188; Falconer 1857, 314; Lyell 1863, 437; letters from H Falconer 20.4.63 & to H Falconer 22.4.63; *OS6*, 302 (*Mesotherium* appears under the name *Typotherium*). **203** <u>The covert evolutionist</u>: F Darwin 1909; Eldredge, 2015; *JR45*, 52 & 173. **203–5** <u>Barnacles</u>: Stott 2004; Richmond 2007; Browne 2003/1, 484; Darwin 1851, 7 & 1854, 3; Herbert 2005, 332. **206–208** <u>Fossils & the Origin</u>: Desmond & Moore 1991, 471–6; *OS*, 301, 305 & 310; *EN*, 135; Eldredge 2015, 173; *OS6*, 275 & 278; *OS* 303, 329 & 339. **208–10** <u>Archaeopteryx</u>: Owen 1863; letters from H Falconer 3.1.63 & to H Falconer 5–6.1.63, to J Dana 7.1.63; *OS4*, 367; *OS5*, 402–3; Switek 2010; Xu 2006. **210** <u>Fossil intermediates</u>: *OS6*, 302 (*Zeuglodon* is now named *Basilosaurus*), 313; *OS*, 343. **210–12** <u>Human fossils</u>: Darwin 1871, 199–201 & 146; Wood 2011; letter to J Hooker 1.9.64; Huxley 1863, 181–4. **212** <u>Responses to the Origin</u>: letter to A Murray 28.4.60; Herbert 2005, 333; Phillips 1860; Allmon 2016; *EN*, 6e; *OS4*, 132. **213** <u>Darwin's associates</u>: Browne 2003/2, 50, 90–4, 117, 122, 131 & 140; Grant 1861; Lyell 1863, 504–6; letter from JV Carus, 15.11.66. **213–4** <u>Owen</u>: Herbert 2005, 325; *BN*, 19 & 161; Rupke 2009, Ch. 5; letter from R Owen, 12.11.59; Owen 1858, 1860; Browne 2003/2, 98 & 110; *OS*, 310 & 329. **214–5** <u>Conclusion</u>: *OS6*, 289; Sepkoski 2013; Gee 2008; Prothero 2007; Asher 2012; Pearson & Ezard 2014.

REFERENCES

Agnolin, F.L. & Chimento, N.R. 2011. Afrotherian affinities for endemic South American "ungulates". *Mamm. Biol.* 76: 101–108.

Aguirre-Urreta, B. & Vennari, V. 2009. On Darwin's footsteps across the Andes: Tithonian-Neocomian fossil invertebrates from the Piuquenes Pass. *Rev. Asoc. Geol. Argentina* 64: 32–42.

Alberdi, M.T. et al. 1995. Patterns of body size changes in fossil and living Equini (Perissodactyla). *Biol. J. Linn. Soc.* 54: 349–370.

Aldiss, D. T. & Edwards, E. J. 1999. The geology of the Falkland Islands. *Br. Geol. Surv. Tech. Rep.* WC/99/10. 135 pp.

Alexander, R. et al.1999. Tail blow energy and carapace fractures in a large glyptodont (Mammalia, Edentata). *Zool. J. Linn. Soc.* 125: 41–49.

Allmon, W.D. 2016. Darwin and palaeontology: a re-evaluation of his interpretation of the fossil record. *Hist. Biol.* 28: 680–706

Ameghino, F. 1889. Contribución al conocimiento de los mamíferos fósiles de la República Argentina. *Actas Acad. Nac. Cienc. Córdoba* 6: 1–98.

Anderson, L. 2009. Fossil trees: Darwin's observations on geographical distribution on either side of the Cordillera. In: Pearn, A. (ed.), 84–5.

Anonymous. 1835. Geological Society. *The Athaneum* 421 (21st November 1835): 876.

Argot, C. 2008. Changing Views in Paleontology: the story of a giant (*Megatherium*, Xenarthra). In: Sargis, E.J. & Dagosto, M. (eds.) *Mammalian Evolutionary Morphology: a Tribute to Frederick S. Szalay*, 37–50. Springer.

Armstrong. P. 1991. *Under the Blue Vault of Heaven: A Study of Charles Darwin's Sojourn in the Cocos (Keeling) Islands.* Nedlands, W.A.: Indian Ocean Centre for Peace Studies.

Armstrong, P. 1992. *Darwin's Desolate Islands: A Naturalist in the Falklands, 1833 and 1834.* Chippenham: Picton.

Asevedo, L. et al. 2012. Ancient diet of the Pleistocene gomphothere *Notiomastodon platensis* from lowland mid-latitudes of South America: Stereomicrowear and tooth calculus analyses combined. *Quat. Int.* 255: 42–52.

Asher, R. 2012. *Evolution and Belief.* CUP.

Banks, M. 1971. A Darwin manuscript on Hobart Town. *Pap. Proc. R. Soc. Tasmania* **105: 5–19.**

Bargo, M.S. & Vizcaíno, S.F. 2008. Paleobiology of Pleistocene ground sloths (Xenarthra, Tardigrada): biomechanics, morphogeometry and ecomorphology applied to the masticatory apparatus. *Ameghiniana* 45: 175–196.

Barlow, N. 1958. *The autobiography of Charles Darwin 1809–1882.* London: Collins. [Darwin 1876]

Barlow, N. (ed.) 1963. Darwin's ornithological notes. *Bull. Brit. Mus. (Nat. Hist.) Hist. Ser.* 2: 201–278. [Darwin 1836–7]

Barnosky, A.D. & Lindsey, E.L. 2010. Timing of Quaternary megafaunal extinction in South America in relation to human arrival and climate change. *Quat. Int.* 217: 10–29.

Barnosky, A.D. et al. 2016. Variable impact of late-Quaternary megafaunal extinction in causing ecological state shifts in North and South America. *PNAS* 113: 856–861.

Barnett, R. and Sylvester, S. 2010. Does the ground sloth, *Mylodon darwinii*, still survive in South America? *Deposits* 23: 8–11.

Barrett, P. H. 1974. The Sedgwick-Darwin geologic tour of North Wales. *Proc. Am. Phil. Soc.* 118: 146–164.

Barrett, P.H. et al. 1987. *Charles Darwin's Notebooks, 1836–1844.* London: British Museum (Natural History).

Beccaloni, G. & Smith, C. 2015. Biography of Wallace. http://wallacefund.info/content/biography-wallace.

Benton, M. J. 1987. Progress and competition in macroevolution. *Biol. Rev.* 62: 305–338.

Blanco, R.E. & Czerowonogora, A. 2003. The gait of *Megatherium* CUVIER 1796. *Senckenbergiana Biol.* 83: 61–8.

Bocherens, H. et al. 2017. Isotopic insight on paleodiet of extinct Pleistocene megafaunal Xenarthrans from Argentina. *Gondwana Res.* 48: 7–14.

Bond, M. 1999. Quaternary native ungulates of Southern South America: a synthesis. *Quaternary of South America and Antarctic Peninsula* 20: 177–205.

Bonney, T.G. 1904. *The Atoll of Funafuti. Borings into a Coral Reef and the Results.* London: Royal Society.

Brea, M. 1997. Una nueva especie del género *Araucarioxylon* Kraus 1870, emend. Maheshwari 1972 del Triásico de Agua de la Zorra, Uspallata. Mendoza. Argentina. *Ameghiniana* 34: 485–496.

Brea, M. et al. 2009. Darwin Forest at Agua de la Zorra: the first *in situ* forest discovered in South America by Darwin in 1835. *Rev. Asoc. Geol. Argentina* 64: 21 – 31.

Brinkman, P.D. 2003. Bartholomew James Sulivan's discovery of fossil vertebrates in the Tertiary beds of Patagonia. *Arch. Nat. Hist.* 30: 56–74.

Brinkman, P.D. 2010. Charles Darwin's Beagle Voyage, Fossil Vertebrate Succession, and "The Gradual Birth & Death of Species" *J. Hist. Biol.* 43: 363–399.

Browne, J. 2003/1. *Charles Darwin: Voyaging.* London: Pimlico.

Browne, J. 2003/2. *Charles Darwin: The Power of Place.* London: Pimlico.

Browne, J. 2008. Introduction. In: Burkhardt (ed.) 2008, ix–xxv.

Buckland, W. 1837. On the adaptation of the structure of the sloths to their peculiar mode of life. *Trans. Linn. Soc.* 17: 17–27.

Buckley, M. et al. 2015. Collagen sequence analysis of the extinct giant ground sloths *Lestodon* and *Megatherium. PLoS ONE* 10(11): e0139611. doi:10.1371/journal. pone.0139611.

Burkhardt, F. 2008. Charles Darwin: *The Beagle Letters.* CUP.

Burkhardt, F. et al. 1985– *The Correspondence of Charles Darwin.* CUP.

Casadío, S. & Griffin, M. 2009. Sedimentology and paleontology of a Miocene marine succession first noticed by Darwin at Puerto Deseado (Port Desire). *Rev. Asoc. Geol. Argentina* 64: 83–89.

Chancellor, G. n.d. Darwin's *Geological diary* from the voyage of the *Beagle.* http://darwin-online.org.uk/ EditorialIntroductions/Chancellor_ GeologicalDiary.html

Chancellor, G. & van Whye, J. 2009. *Charles Darwin's Notebooks from the Voyage of the Beagle.* CUP.

Charrier, R. et al. 2006. *Tectonostratigraphic evolution of the Andean Orogen in Chile.* In: Moreno, T. & *Gibbons, W.* (eds). *Geology of Chile, 21–114.* London: Geological Society.

Clift, W. 1832–1837. *Diaries.* Manuscripts, Royal College of Surgeons, London.

Clift, W. 1835. Notice on the Megatherium brought from Buenos Ayres by Woodbine Parish. *Trans. Geol. Soc. Lond.* (2nd ser.) 3: 437–450.

Cope, E.D. 1891. The Litopterna. *Am. Nat.* 25: 685–693.

Cuvier, G. 1796. Notice sur le squelette… *Magasin Encyclopédique* 1: 303–310.

Cuvier, G. 1806. Sur le grand mastodonte. *Ann. Mus. d'Hist. Nat.* 8: 270–312.

Cuvier. G. 1812. Sur le *Megatherium….* In: *Recherches sur les Ossemens Fossiles,* vol IV, part IV, 19–43.

Daly, R. A. 1915. The glacial-control theory of coral reefs. *Proc. Am. Acad. Arts Sci.* 51: 155–251.

Dantas, M.A.T. et al. 2013. A review of the time scale and potential geographic distribution of *Notiomastodon platensis* (Ameghino, 1888) in the late Pleistocene of South America. *Quat. Int.* 317: 73–79.

Darwin, C. 1832–33. *Beagle animal notes* (ed. R.D. Keynes). CUL-DAR29.1.A1–A49.

Darwin, C. 1832-36. *Geological Diary.* CUL-DAR.32–38.

Darwin, C. 1834. *Reflection on reading my geological notes.* CUL-DAR42.93–6 & 42.148.

Darwin, C. 1835a. *Coral Islands.* CUL-DAR41.1–12.

Darwin, C. 1835b. *Extracts from letters addressed to Professor Henslow.* Privately printed, Cambridge Philosophical Society, Dec. 1835.

Darwin, C. 1835c. *The position of the bones of Mastodon (?) at Port St Julian is of interest…* CUL-DAR42.97–99.

Darwin, C. 1836. *Notes on the Geology and Corals of the Keeling Islands.* CUL-DAR41.40–57.

Darwin, C. 1836–7. *Ornithological Notes* (ed. N. Barlow). *Bull. Brit. Mus. (Nat. Hist.) Hist. Ser.* 2: 201–278 (1963).

Darwin, C. 1839. *Journal and Remarks.* London: Henry Colburn.

Darwin, C. 1840. On the connexion of certain volcanic phenomena in South America; and on the formation of mountain chains and volcanos, as the effect of the same powers by which continents are elevated. *Trans. Geol. Soc. Lond.* (Ser. 2) 5: 601–631.

Darwin, C. 1842. *The Structure and Distribution of Coral Reefs.* London: Smith, Elder.

Darwin, C. 1844. *Geological Observations on the Volcanic Islands Visited During the Voyage of H.M.S. Beagle.* London: Smith, Elder.

Darwin, C. 1845. *Journal of Researches* (2nd edition of Darwin 1839). London: John Murray.

Darwin, C. 1846. *Geological Observations on South America.* London: Smith, Elder.

Darwin, C. 1851. *Fossil Cirripedia of Great Britain: A monograph on the fossil Lepadidae, or pedunculated cirripedes of Great Britain.* London: Palaeontographical Society.

Darwin, C. 1854. *A monograph on the fossil Balanidæ and Verrucidæ of Great Britain.* London: Palaeontographical Society.

Darwin, C. 1859. *On the origin of species by means of natural selection, or the preservation of favoured races in the struggle for life* [1st edition]. 2nd edn, 1860; 3rd edn, 1861; 4th edn, 1866; 5th edn, 1869; 6th edn, 1872. London: John Murray.

Darwin, C. 1871. *The Descent of Man, and Selection in Relation to Sex.* London: John Murray.

Darwin, C. 1876. *The Autobiography of Charles Darwin 1809–1882.* (Ed. N. Barlow, 1958). London: Collins.

Darwin, F, ed. 1909. *The foundations of The origin of species. Two essays written in 1842 and 1844.* Cambridge: CUP.

Dawson, G. 2016. *Show Me The Bone.* Chicago UP.

Delsuc, F. et al. 2016. The phylogenetic affinities of the extinct glyptodonts. *Curr. Biol.* 26: R155–R156.

Deschamps, C.M. et al. 2012. Biostratigraphy and correlation of the Monte Hermoso Formation (early Pliocene, Argentina): The evidence from caviomorph rodents. *J. South Am. Earth Sci.* 35: 1–9.

Desmond, A. & Moore, J. 1991. *Darwin.* London: Michael Joseph.

Dobbs, D. 2005. *Reef Madness.* NY: Pantheon.

D'Orbigny, A. 1842. *Voyage dans l'Amérique Méridionale,* Vol. 3, Part 4 (Paléontologie). Paris.

Dumitrica, P. 2007. Phytolitharia. *DSDP* 13: 940–943. http://www. deepseadrilling.org/13/volume/ dsdp13pt2_34_4.pdf.

Ehrenberg, C.G. 1845. Vorläufige zweite Mittheilungen über die weiterer Erkenntnis der Beziehungen des kleinsten organischen Lebens zu den vulkanischen Massen der Erde. *Bericht Bekannt. Verhand. Konigl. Preuss. Akad. Wiss. Berlin,* April 1845, 133–158.

Ehrenberg, C.G. 1854. *Mikrogeologie.* 2 vols. Leipzig.

Eldredge, N. 2015. *Eternal Ephemera.* NY: Columbia UP.

Elissamburu, A. 2004. Análisis morfométrico y morfofuncional del esqueleto apendicular de *Paedotherium* (Mammalia, Notoungulata). *Ameghiniana* 41: 363–380.

Encinas, A. et al. 2014. Comment on Reply to Comment of Finger *et al.* (2013) on: 'Evidence for an Early-Middle Miocene age of the Navidad Formation (central Chile)…'. *Andean Geology 41: 639–656.*

Falcon-Lang, H. 2012. Fossil 'treasure trove' found in British Geological Survey vaults. *Geology Today* 28: 26–30.

Falcon-Lang, H. J. & Digrius, D.M. 2014. Palaeobotany under the microscope: history of the invention and widespread adoption of the petrographic thin section technique. *Quekett J. Microsc.* 42: 253–280.

Falconer, H. 1857. On the species of mastodon and elephant occurring in the fossil state in Great Britain. Part I. Mastodon. *Quart. J. Geol. Soc.* 13: 314.

Falkner, T. 1774. *A description of Patagonia, and the adjoining parts of South America.* Hereford: C. Pugh.

Fariña, R.A., Blanco R.E. & Christiansen, P. 2005. Swerving as the escape strategy of *Macrauchenia patachonica* (Mammalia; Litopterna). *Ameghiniana* 42: 751–760.

Fariña, R.A., Vizcaíno, S.F., de Iuliis, G. 2013. *Megafauna: Giant Beasts of Pleistocene South America.* Bloomington: Indiana UP.

Farinati, E.A., 1985. Radiocarbon dating of Holocene marine deposits, Bahía Blanca area, Buenos Aires Province, Argentina. *Quaternary of South America and Antarctic Peninsula* 3: 197–206.

Fernández, M.E. et al 2000. Functional morphology and palaeobiology of the Pliocene rodent *Actenomys* (Caviomorpha: Octodontidae): the evolution to a subterranean mode of life. *Biol. J. Linn. Soc.* 71: 71–90.

Fernicola, J.C. et al. 2009. The fossil mammals collected by Charles Darwin in South America during his travels on board the HMS Beagle. *Rev. Asoc. Geol. Argentina* 64: 147 – 159.

Ferretti, M.P. 2011. Anatomy of *Haplomastodon chimborazi* (Mammalia, Proboscidea) from the late Pleistocene of Ecuador and its bearing on the phylogeny and systematics of South American gomphotheres. *Geodiversitas* 32: 663–721.

Fitzroy, R. 1839. *Narrative of the surveying voyages of His Majesty's Ships Adventure and Beagle between the years 1826 and 1836… Proceedings of the second expedition, 1831–36…* London: Henry Colburn.

França, L.M. et al. 2015. Review of feeding ecology data of Late Pleistocene mammalian herbivores from South America and discussions on niche differentiation. *Earth-Science Rev.* 140: 158–165.

Gee, H. 2008. *Deep Time: Cladistics, the Revolution in Evolution.* London: HarperCollins.

Glynn, P.W. et al. 2015. Coral reef recovery in the Galápagos Islands: the northernmost islands (Darwin and Wenman). *Coral Reefs* 34: 421–436.

Grant, R.E. 1861. *Tabular view of the primary divisions of the animal kingdom…* London: Royal College of Surgeons.

Grayson, D.K. 1984. Nineteenth-century explanations of Pleistocene extinctions: a review and analysis. In: Martin, P.S. & Klein, R.G. (eds.) *Quaternary Extinctions: A Prehistoric Revolution,* 5–39. Tucson: University of Arizona Press.

Griffin, M. & Nielsen, S.N. 2008. A revision of the type specimens of Tertiary molluscs from Chile and Argentina described by d'Orbigny (1842), Sowerby (1846) and Hupé (1854). *J. Syst. Pal.* 6: 251–316.

Griffith, E. 1827–35. *The Animal Kingdom* [English translation, with additions, of Cuvier's *Règne Animal*]

Grigg, R.W. 2011. Darwin Point. In: Hopley, D. (ed) *Encyclopedia of Modern Coral Reefs,* 298–299. Dordrecht: Springer.

Groombridge, M. (ed.) 1992. *Global Biodiversity: Status of the Earth's Living Resources.* London: Chapman & Hall.

Gross, M. et al. 2015. A minute ostracod (Crustacea: Cytheromatidae) from the Miocene Solimões Formation (western Amazonia, Brazil): evidence for marine incursions? *J. Syst. Pal.* 14: 581–602.

Hall, R. 2012. Late Jurassic–Cenozoic reconstructions of the Indonesian region and the Indian Ocean. *Tectonophysics* 570–1: 1–41.

Heads, M. 2006. Panbiogeography of *Nothofagus* (Nothofagaceae): analysis of the main species massings. *J. Biogeogr.* 33: 1066–1075.

Herbert, S. 2005. *Charles Darwin, Geologist.* Ithaca: Cornell UP.

Hoernle, K. et al. 2011. Origin of Indian Ocean Seamount Province by shallow recycling of continental lithosphere. *Nat. Geosc.* 4: 883–7.

Hutton, A. C. 2009. Geological setting of Australasian coal deposits. In R. Kininmonth & E. Baafi (eds.), *Australasian Coal Mining Practice,* 40–84. The Australasian Institute of Mining and Metallurgy.

Huxley, T.H. 1863. *Evidence as to Man's Place in Nature.* NY: Appleton.

Iriondo, M. & Krohling, D. 2009. From Buenos Aires to Santa Fe: Darwin's observations and modern knowledge. *Rev. Asoc. Geol. Argentina* 64: 109 – 123.

IUCN, 2017. *Chilonopsis nonpareil.* http://www.iucnredlist.org/details/4639/0.

Jameson, R. 1827. Of the changes which life has experienced on the globe. *Edinb. New Philos. J.* 3: 298–301. [published anonymously]

Johnson, M.E. et al. 2012. Rhodoliths, uniformitarianism, and Darwin: Pleistocene and Recent carbonate deposits in the Cape Verde and Canary archipelagos. *Palaeo-3* 329–330: 83–100.

Jordan, G.J. & Hill, R.S. 2002. Cenozoic plant macrofossil sites of Tasmania. *Pap. Proc. R. Soc. Tasmania* 136: 127–139.

Judd, J.W. 1890. Critical Introduction. In: Darwin, C. *On the structure and distribution of coral reefs*, 3–10. London: Ward, Lock.

Judd, J.W. 1911. Charles Darwin's earliest doubts concerning the immutability of species. *Nature* 88: 8–12.

Kennedy, W.J. & Henderson, R.A. 1992. Heteromorph ammonites from the Upper Maastrichtian of Pondicherry, south India. *Palaeontology* 35: 693–731.

Kenrick, P. & Davis, P. 2004. *Fossil Plants*. London: Natural History Museum.

Keynes, R.D. (ed.) 1988, *Charles Darwin's Beagle Diary*. CUP.

Keynes, R. D. (ed.) 2000. *Charles Darwin's zoology notes & specimen lists from H.M.S. Beagle*. CUP.

Kious, W.J. & Tilling, R.I. 1996. *This Dynamic Earth: The Story of Plate Tectonics*. Diane Publishing.

Knapp, M. et al. 2005. Relaxed molecular clock provides evidence for long-distance dispersal of *Nothofagus* (southern beech). *PLoS Biol.* 3(1): e14.

Kohn, D. et al. 2005. What Henslow taught Darwin. *Nature* 465: 643–5.

Lambert, J. & Jeannet, A. 1928. Nouveau catalogue des moules d'échinides fossiles du MHN. Exécutés sous la direction de L. Agassiz et E. Desor. *Mem. Soc. Helv. Sc. Nat.* 64/2: 1–233.

Larramendi, A. 2016. Shoulder height, body mass, and shape of proboscideans. *Acta Pal. Pol.* 61: 537–574.

Li, Y. & Han, W. 2015. Decadal sea level variations in the Indian Ocean investigated with HYCOM: Roles of climate modes, ocean internal variability, and stochastic wind forcing. *J. Climate* 28: 9143–65.

Lucas, S.G. 2013. The palaeobiogeography of South American gomphotheres. *J. Palaeogeog.* 2: 19–40.

Lyell, C. 1830–33. *Principles of Geology*, vols. 1–3. London: John Murray.

Lyell, C. 1833–38a. Address to the Geological Society, 19th February 1836. *Proc. Geol. Soc. Lond.* 2: 357–390.

Lyell, C. 1833–38b. Address to the Geological Society, 17th February

1837. *Proc. Geol. Soc. Lond.* 2: 479–523.

Lyell, C. 1844. On the geological position of the *Mastodon giganteum* and associated fossil remains at Bigbone Lick, Ky., and other localities in the United States and Canada. *Am. J. Sci.* 46: 320–323.

Lyell, C. 1863. *Geological Evidences of the Antiquity of Man* (3rd edn). London: John Murray.

Masse, J.-P. et al. 2015. Aptian-Albian rudist bivalves (Hippuritida) from the Chilean Central Andes: their palaeoceanographic significance. *Cret. Res.* 54: 243–254.

Metcalf, J.L. et al. 2016. Synergistic roles of climate warming and human occupation in Patagonian megafaunal extinctions during the Last Deglaciation. *Sci. Adv.* 2: e1501682.

Mishra, D.P. n.d. *Spontaneous Combustion*. http://www.slideshare.net/mj2611/spontaneous-combustion-of-coal

Moore, D. M. 1978. Post-glacial vegetation in the South Patagonian territory of the giant ground sloth, *Mylodon*. *Bot. J. Linn. Soc.* 77: 177–202.

Morgan, C.C. & Verzi, D.H. 2011. Carpal-metacarpal specializations for burrowing in South American octodontoid rodents. *J. Anat.* 219: 167–75.

Morris, J. 1845. *Descriptions of fossils, Mollusca. In* P.E. de Strzelecki (ed.) *Physical Descriptions of New South Wales and van Diemen's Land*, 270–290. London: Longman.

Morris, J. & Sharpe, D. 1846. Description of eight species of brachiopodous shells from the Palaeozoic rocks of the Falkland Islands. *Proc. Geol. Soc. Lond.* 2: 274–8.

Mothé, D. et al. 2012. Taxonomic revision of the Quaternary gomphotheres (Mammalia: Proboscidea: Gomphotheriidae) from the South American Lowlands. *Quat. Int.* 276–277: 2–7.

Mothé, D. et al. 2016. Sixty years after 'The mastodonts of Brazil': The state of the art of South American proboscideans (Proboscidea, Gomphotheriidae). *Quat. Int.* 443A: 52–64.

Mothé, D. et al. 2017. Early humans and South American proboscideans: What do the paleoarchaeological sites reveal? Abstracts, VII ICMR, Taichung 17–23 Sept 2017, HI1–5.

Murchison, R.I. 1839. *The Silurian System*. London: John Murray.

Nielsen, S.N. & Salazar, C. 2011. *Eutrephoceras subplicatum* (Steinmann, 1895) is a junior synonym of *Eutrephoceras dorbignyanum* (Forbes in Darwin, 1846) (Cephalopoda, Nautiloidea) from the Maastrichtian Quiriquina Formation of Chile. *Cretac. Res.* 32: 833–840.

Olivero, E.B. et al. 2009. The stratigraphy of Cretaceous mudstones in the eastern Fuegian Andes: new data from body and trace fossils. *Rev. Asoc. Geol. Argentina* 64: 60–69.

Orlando, L. et al. 2008. Ancient DNA clarifies the evolutionary history of American Late Pleistocene equids. *J. Mol. Evol.* 66: 533–8.

Osborn, H.F. 1936. *Proboscidea*, vol. 1. New York: AMNH.

Owen, R. 1837. A description of the Cranium of the Toxodon Platensis, a gigantic extinct mammiferous species, referrible by its dentition to the Rodentia, but with affinities to the Pachydermata and the Herbivorous Cetacea. *Proc. Geol. Soc. Lond.* 2: 541–2.

Owen, R. 1838–1840. *Zoology of the Voyage of H.M.S. Beagle* (in four parts). London: Smith, Elder.

Owen, R. 1840–45. *Odontography*. London: Hippolyte Bailliere.

Owen, R. 1841. Description of a tooth and part of the skeleton of the *Glyptodon clavipes…* *Trans. Geol. Soc. Lond.*, Ser. 2, 6: 81–106.

Owen, R. 1842. *Description of the skeleton of an extinct gigantic sloth, Mylodon robustus Owen, with observations on the osteology, natural affinities, and probable habits of the megatherioid quadrupeds in general*. London: Taylor.

Owen, R. 1845. *Descriptive and Illustrated Catalogue of the Fossil Organic Remains of Mammalia and Aves Contained in the Museum of the Royal College of Surgeons of England*. London: Taylor.

Owen, R. 1846. Notices of some fossil Mammalia of South America. *Report of the 16th meeting of the British Association for the Advancement of Science, Notices and Abstracts:* 65–67.

Owen R. 1851–9. On the *Megatherium* (*Megatherium americanum*, Blumenbach), Parts I–V. *Transactions of the Royal Society of London* 141: 719–764, 145: 359–388, 146: 571–589, 148: 261–278, 149: 809–829.

Owen, R. 1853. Description of some species of the extinct genus *Nesodon*, with remarks on the primary group (Toxodontia) of hoofed quadrupeds, to which that genus is referable. *Phil. Trans. R. Soc. Lond.* 143: 291–310.

Owen, R. 1858. Address. *Br. Ass. Adv. Sci., Rep.*, xlix–cx.

Owen, R. 1860. Darwin on the Origin of Species. *Edinburgh Review* 3: 487–532.

Owen, R. 1863. On the *Archaeopteryx* of von Meyer. *Phil. Trans. R. Soc. Lond.* 153: 33–46.

Pander, C.H. & d'Alton, E.J. 1821. Das Riesen-Faulthier, *Bradypus giganteus*... In: Pander, C.H. & d'Alton, E.J. (eds.) *Vergleichende Osteologie*, vol. 1 part 1. Bonn: Weber.

Pant, S.R. et al. 2014. Complex body size trends in the evolution of sloths (Xenarthra: Pilosa). *BMC Evol. Biol.* 14: 184.

Parish, W. 1839. *Buenos Ayres, and the Provinces of the Rio de la Plata.* London: John Murray.

Parodiz, J.J. 1981. *Darwin in the New World.* Leiden: Brill.

Parras, A. & Griffin, M. 2009. Darwin's great Patagonian Tertiary Formation at the mouth of the Rio Santa Cruz: a reappraisal. *Rev. Asoc. Geol. Argentina* 64: 70–82.

Pearn, A.E. (Ed.). 2009. *A Voyage Around the World.* CUP.

Pearson, P.N. & Ezard, T.H.G. 2014. Evolution and speciation in the Eocene planktonic foraminifer *Turborotalia. Paleobiology* 40: 130–143.

Pearson, P.N. & Nicholas, C.J. 2007. 'Marks of extreme violence': Charles Darwin's geological observations at St Jago (São Tiago),

Cape Verde islands *Spec. Publ. Geol. Soc. Lond.* 287: 239–253.

Pedoja, K. et al. 2011. Uplift of Quaternary shorelines in eastern Patagonia: Darwin revisited. *Geomorphology* 127: 121–142.

Perry, C.T. et al. 2015. Linking reef ecology to island building: Parrotfish identified as major producers of island-building sediment in the Maldives. *Geology* 43: 503–6.

Philippe, M. et al. 1998. Tertiary and Quaternary fossil wood from Kerguelen (southern Indian Ocean). *C. R. Acad. Sci., Sci. terre planètes* 326: 901–906.

Phillips, J. 1860. *Life on the Earth: its Origin and Succession.* Cambridge: Macmillan.

Pick, N. 2004. *The rarest of the rare. Stories behind the treasures at the Harvard Museum of Natural History.* NY: HarperCollins.

Poma, S. et al. 2009. Darwin's observation in South America: what did he find at Agua de la Zorra, Mendoza Province? *Rev. Asoc. Geol. Argentina* 64: 13 – 20.

Porter, D.M. 1985. The *Beagle* collector and his collections. In: Kohn, D. ed. *The Darwinian Heritage*, 973–1019. Princeton UP.

Prado, J.L. et al. 2011. Ancient feeding ecology inferred from stable isotopic evidence from fossil horses in South America over the past 3 Ma. *BMC Ecology* 2011, 11:15.

Prado, J.L. & Alberdi, M.T. 2014. Global evolution of Equidae and Gomphotheriidae from South America. *Integr. Zool.* 9: 434–44

Prado, J.L. et al. 2015. Megafauna extinction in South America: A new chronology for the Argentine Pampas. *Palaeo-3* 425: 41–49.

Prothero, D. 2007. *Evolution: What the fossils say and why it matters.* NY: Columbia UP.

Quattrocchio, M.E. et al. 2009. Geology of the area of Bahía Blanca, Darwin's view and the present knowledge: a story of 10 million years. *Rev. Asoc. Geol. Argentina* 64: 137–146.

Quoy, J. R. C. & Gaimard, J. P. 1824. Histoire Naturelle: Zoologie. In: Freycinet, L. de. *Voyage autour du monde.* Paris: Imprimerie Royale.

Rachootin, S. 1985. Owen and Darwin reading a fossil: Macrauchenia in a boney light. In: Kohn, D. (ed.) *The Darwinian Heritage*, 155–183. Princeton UP. Richmond, M. 2007. Darwin's Study of the Cirripedia. http://darwin-online.org.uk/EditorialIntroductions/Richmond_cirripedia.html.

Roberts, M.B. 2001. Just before the Beagle: Charles Darwin's geological fieldwork in Wales, summer 1831. *Endeavour* 25: 33–7.

Rößler, R. et al. 2014. Which name(s) should be used for *Araucaria*-like fossil wood? Results of a poll. *Taxon* 63: 177–184.

Rosen, B.R. 1982. Darwin, coral reefs, and global geology. *Bioscience* 32: 519–525.

Rosen, B.R. & Darrell, J. 2010. A generalised historical trajectory for Charles Darwin's specimen collections, with a case study of his coral reef specimen list in the Natural History Museum, London. In: Stoppa, F. & Veraldi, R. (eds.) *Darwin tra Storia e Scienza*, 133–194. Edizioni Universitarie Romane.

Rosen, B.R. & Darrell, J. 2011. Darwin, pioneer of reef transects, reef ecology and ancient reef modelling: significance of his specimens in the Natural History Museum, London. *Kölner Forum Geol. Paläont.* 19: 151–3.

Rostami, K. et al. 2000. Quaternary marine terraces, sea-level changes and uplift history of Patagonia, Argentina: comparisons with predictions of the ICE-4G (VM2) model of the global process of glacial isostatic adjustment. *Quat. Sci. Rev.* 19: 1495–1525.

Rupke, N. 2009. *Richard Owen.* Chicago UP.

Saarinen, J. & Karme, A. 2017. Tooth wear and diets of extant and fossil xenarthrans (Mammalia, Xenarthra) – applying a new mesowear approach. *Palaeo-3* 476: 42–54.

Schellman, G. & Radtke, U. 2010. Timing and magnitude of Holocene sea-level changes along the middle and south Patagonian Atlantic coast derived from beach ridge systems, littoral terraces and valley-mouth terraces. *Earth–Science Rev.* 103: 1–30.

Scott, G.A.J. & Rotondo, G.M. 1983. A model to explain the differences between Pacific plate island atoll types. *Coral Reefs* 1: 139–150.

Sepkoski, D. 2013. Evolutionary paleontology. In: Ruse, M. (ed.) *The Cambridge Encyclopedia of Darwin and Evolutionary Thought*, 353–360. CUP.

Shockey, B.J. 2001. Specialized knee joints in some extinct, endemic, South American herbivores. *Acta Pal. Pol.* 46: 277–288.

Simpson, G. G. 1984. *Discoverers of the Lost World.* New Haven: Yale UP.

Slater, G. et al. 2016. Evolutionary relationships among extinct and extant sloths: the evidence of mitogenomes and retroviruses. *Genome Biol. Evol.* 8: 607–621.

Sponsel, A. 2016. An amphibious being: how maritime surveying reshaped Darwin's approach to natural history. *Isis* 107: 254–281.

Steers, J.A. & Stoddart, D.R. 1977. The origin of fringing reefs, barrier reefs, and atolls. In: Jones, O.A. & Endean, R. (eds.) *Biology and Geology of Coral Reefs* vol. IV: Geology 2, 21–57. New York: Academic Press.

Stoddart, D.R. 1976. Darwin, Lyell, and the geological significance of coral reefs. *Br. J. Hist. Sci.* 9: 199–218.

Stoddart, D.R. 1995. Darwin and the seeing eye. *Earth Sci. Hist.* 14: 3–22.

Stone, P. & Rushton, A.W.A. 2012. The pedigree and influence of fossil collections from the Falkland Islands: From Charles Darwin to continental drift. *PGA* 123: 520–532.

Stone, P. & Rushton, A.W.A. 2013. Charles Darwin, Bartholomew Sulivan, and the geology of the Falkland Islands: unfinished business from an asymmetric partnership. *Earth Sci. Hist.* 32: 156–185.

Stone, P. et al. 2015. Charles Darwin's 'Gorgonia' – a palaeontological mystery from the Falkland Islands resolved. *Falkland Is. J.* 10: 6–15.

Stott, R. 2004. *Darwin and the Barnacle.* NY: Norton.

Stott, R. 2012. *Darwin's Ghost: In Search of the First Evolutionists.* London: Bloomsbury.

Sulloway, F. J. 1982. Darwin's conversion: The *Beagle* voyage and its aftermath. *J. Hist. Biol.* 15: 325–396.

Switek, B. 2010. Thomas Henry Huxley and the reptile to bird transition. *Geol. Soc. Spec. Pubs.* 343: 251–263.

Thomas, B.A. 2009. Darwin and plant fossils. *The Linnean* 25/2: 24–42.

Toledano, A.M. 2011. *The Posthumous Lives of the Giant Sloth: The Megatherium's Path from Artifact to Idea.* Thesis, Princeton, NJ, http://arks.princeton.edu/ark:/88435/dsp01cf95jc75j.

Varela, L. & Fariña, R.A. 2016. Co-occurrence of mylodontid sloths and insights on their potential distributions during the late Pleistocene. *Quat. Res.* 85: 66–74.

Viens, R. 2014. Ammonites on top of Mount Tarn. https://beagleproject.wordpress.com/2014/02/16/ammonites-on-top-of-mount-tarn/.

Vizcaíno, S.F. et al. 2010. Proportions and function of the limbs of glyptodonts. *Lethaia* 44: 93–101.

Vizcaíno, S.F. et al. 2011. Evaluating habitats and feeding habits through ecomorphological features in Glyptodonts (Mammalia, Xenarthra). *Ameghiniana* 48: 305–319.

Vucetich, M.G. et al. 2013. Paleontology, evolution and systematics of capybara. In (Moreira, J.R. et al., eds.) *Capybara.* NY: Springer.

Welker, F. et al. 2015. Ancient proteins resolve the evolutionary history of Darwin's South American ungulates. *Nature* 522: 81–4.

Wesson, R. 2017. *Darwin's First Theory.* NY: Pegasus.

Williams, S. 2017. Molluscan shell colour. *Biol. Rev.* 92: 1039–1058.

Wilson, L.G. 1972. *Charles Lyell. The Years to 1841: The Revolution in Geology.* New Haven: Yale UP.

Winslow, J.H. 1975. Mr Lumb and Masters Megatherium: an unpublished letter by Charles Darwin from the Falklands. *J. Hist. Geog.* 1: 347–360.

Wood, B. (ed.) 2011. *Wiley-Blackwell Encyclopedia of Human Evolution.* Wiley-Blackwell.

Woodroffe, C.D. 2005. Late Quaternary sea-level highstands in the central and eastern Indian Ocean: a review. *Global & Planetary Change* 49: 121–138.

Woodroffe, C.D. et al. 1990. Sea level and coral atolls: Late Holocene emergence in the Indian Ocean. *Geology* 18: 62–66.

Woodroffe, C.D. et al. 1991. Last interglacial reef and subsidence of the Cocos (Keeling) Islands, Indian Ocean. *Marine Geol.* 96: 137–143.

Woodroffe, C.D. & MacLean, R.F. 1994. Reef islands of the Cocos (Keeling) Islands. *Atoll Res. Bull.* 402: 1–36.

Wyse Jackson, P.N. et al. 2011. The status of *Protoretepora* de Koninck, 1878 (Fenestrata: Bryozoa), and description of *P. crockfordae* sp. nov. and *P. wassi* sp. nov. from the Permian of Australia. *Alcheringa* 35: 539–552.

Xu, X. 2006. Scales, feathers and dinosaurs. *Nature* 440: 287–8.

Zárate, M. & Folguera, A. 2009. On the formations of the Pampas in the footsteps of Darwin: south of the Salado. *Rev. Asoc. Geol. Argentina* 64: 124–136.

Zurita, A. E. et al. 2008. A new species of *Neosclerocalyptus* Paula Couto, 1957 (Xenarthra, Glyptodontidae, Hoplophorinae) from the middle Pleistocene of the Pampean region, Argentina. *Geodiversitas* 30: 779–791.

Zurita, A. E. et al. 2010. Accessory protection structures in *Glyptodon* Owen (Xenarthra, Cingulata, Glyptodontidae). *Ann. Pal.* 96: 1–11.

Zurita, A. E. et al. 2011. *Neosclerocalyptus* spp. (Cingulata: Glyptodontidae: Hoplophorini): cranial morphology and palaeoenvironments along the changing Quaternary. *J. Nat. Hist.* 45:15–16.

FURTHER INFORMATION

ONLINE RESOURCES

John van Wyhe (Ed.). 2002. *The Complete Work of Charles Darwin Online* (http://darwin-online.org.uk/). All of Darwin's published works and most of his manuscripts, including his field notebooks, geological diary, *Beagle* diary, transmutation notebooks, *Autobiography*, *Journal of Researches* and all editions of the *Origin of Species.*

Darwin Correspondence Project (https://www.darwinproject.ac.uk/). A large selection of the letters sent and received by Darwin.

Aguirre-Urreta, B. et al. (eds.) 2009. Darwin's Geological Research in Argentina. *Revista de la Asociación Geológica Argentina*, vol. 64, part 1 (http://darwin-online.org.uk/ converted/pdf/2009_Revista_A194.pdf).

BOOKS

CHAPTER 1 & GENERAL

Browne, J. 2003. *Charles Darwin: Voyaging*, and *Charles Darwin: The Power of Place. Volumes 1 and 2 of a Biography.* London: Pimlico.

Burkhardt, F. (Ed.). 2008. *Charles Darwin: The Beagle Letters.* Cambridge University Press.

Darwin, C. 1845. *The Voyage of the Beagle* (multiple editions).

Keynes, R. 2002. *Fossils, Finches and Fuegians: Charles Darwin's Adventures and Discoveries on the Beagle, 1832-1836.* London: HarperCollins.

Pearn, A.M. (Ed.). 2009. *A Voyage Round the World.* Cambridge University Press.

CHAPTER 2

Fariña, R., Vizcaíno, S.F., Iuliis, G. 2013. *Megafauna: Giant Beasts of Pleistocene South America.* Bloomington: Indiana University Press.

Dawson, G. 2016. *Show Me the Bone: Reconstructing Prehistoric Monsters in Nineteenth-Century Britain and America.* London: University of Chicago Press.

CHAPTER 3

Kenrick, P. & Davis, P. 2004. *Fossil Plants.* London: Natural History Museum.

CHAPTER 4

Herbert, S. 2005. *Charles Darwin, Geologist.* New York: Cornell University Press.

Wesson, R. 2017. *Darwin's First Theory: Exploring Darwin's Quest to Find a Theory of the Earth.* London: Pegasus.

CHAPTER 5

Dobbs, D. 2005. *Reef Madness: Charles Darwin, Alexander Agassiz, and the Meaning of Coral.* New York: Pantheon.

CHAPTER 6

Eldredge, N. 2015. *Eternal Ephemera: Adaptation and the Origin of Species from the Nineteenth Century through Punctuated Equilibria and Beyond.* New York: Columbia University Press.

Prothero, D.R. 2007. *Evolution: What the Fossils Say and Why it Matters.* New York: Columbia University Press.

INDEX

Page numbers in *italic* refer to illustration captions; those in **bold** refer to main subjects of boxed text.

PICTURE SOURCES

LIST OF SPECIMENS ILLUSTRATED

All are Darwin specimens except those in brackets. Abbreviations: BGS - British Geological Survey, Keyworth; BMNH - Natural History Museum, London; CAMSM – Sedgwick Museum, Cambridge; DH – Down House, Kent; MCZ – Museum of Comparative Zoology, Harvard; RCSHM – Royal College of Surgeons, London. All other specimens are Natural History Museum, London and are prefixed NHMUK.

27 wood PB V 5668. **37** *Megatherium* top & bottom: RCSHM/CO 3443; middle: DH. **41** *Megatherium* PV M 16585. **46** *Mylodon* PV M 16563a. **48** *Scelidotherium* PV M 82206. **50** *Glossotherium* PV M 16586. (**57** *Glyptodon* PV M 4473). **60** *Equus* PV M 16558. **76 & 79** *Toxodon* PV M 16560. **77** *Toxodon* PV M 16566. **78** *Toxodon* PV M 16567. **86 & 89** *Macrauchenia* PV M 43402j-o; (tapir 1948.12.20.3; llama GERM 674a). (**92** *Glossopteris* PB V 7350). **94** wood PB V 5141. **95** wood section BGS PF 7455. **96** wood PB V 5136 & 4788. **99** wood PB V 4790. **103** wood PB V 5284 & 5592. **104** wood PB V 5236. **105** lignite CAMSM 112391-2. **107** *Nothofagus* PB V 21578. (**108** *Glossopteris* PB V 7292). (**110** travertine PB V 156). **112** *Crassostrea* PI. **114** rhodoliths CAMSM 111928. **116** *Erodona* CAMSM X.50292. **119** *Balanus* PI OR 38436. **123** *Tegula* CAMSM D.17160-7; *Zidona* CAMSM D.17150; *Astraea* CAMSM D.17205-6. **125** electrid bryozoan CAMSM D.17207-9. **130** *Scelidotherium* PV M 82206d&h. **131** *Zygochlamys* PI L 27960. **132** *Iheringiana* MCZ 2496. **133** *Trophon* PI G 26415; *Hesperibalanus* PI OR 38435; *Adelomelon* PI G 25287. **135** *Fissidentalium* PI G 26395. **137** *Pristis* MCZ 8591. **138** *Incatella* PI G 26418. **140** *Maorites* PI C 2612. **146** *Eubaculites* PI C 2611. **147** *Eutrephoceras* PI C 2613. **149** *Rhynchonella* PI B 14675-6. **151** *Terebratula* PI B 18314 & PI OR 30518. **152** *Gryphaea* PI LL 27655. **155** Falkland brachiopods PI OR 17794. **157** *Bainella* PI B 17790. **159** Falkland crinoids CAMSM 112303. **161** *Terrakea* PI B 19298; *Spirifer* PI B 10858. **169** *Porites* BMNH 1842.12.14.24; *Acropora* BMNH 1842.12.14.3; *Millepora* BMNH 1842.12.14.29; coralline alga BMNH 1842.12.14.30. **170** *Acropora* BMNH 1842.12.14.37 & 1842.12.14.39. **171** reef rock BMNH 1842.12.14.47-50 & 1842.12.14.44; coral pebbles BMNH 1842.12.14.52-65. **187** *Zaedyus* 1855.12.24.288. (**211** *Homo* PA EM 3811).